# SUMP for Cities' Sustainable Development

# SUMP for Cities' Sustainable Development

Editors

**Marija Burinskienė**
**Rasa Ušpalytė-Vitkūnienė**

MDPI • Basel • Beijing • Wuhan • Barcelona • Belgrade • Manchester • Tokyo • Cluj • Tianjin

*Editors*
Marija Burinskienė
Vilnius Gediminas Technical University
Lithuania

Rasa Ušpalytė-Vitkūnienė
Vilnius Gediminas Technical University
Lithuania

*Editorial Office*
MDPI
St. Alban-Anlage 66
4052 Basel, Switzerland

This is a reprint of articles from the Special Issue published online in the open access journal *Sustainability* (ISSN 2071-1050) (available at: https://www.mdpi.com/journal/sustainability/special_issues/sump_sus).

For citation purposes, cite each article independently as indicated on the article page online and as indicated below:

LastName, A.A.; LastName, B.B.; LastName, C.C. Article Title. *Journal Name* **Year**, *Volume Number*, Page Range.

**ISBN 978-3-0365-0460-5 (Hbk)**
**ISBN 978-3-0365-0461-2 (PDF)**

Cover image courtesy of Rasa Ušpalytė.

# Contents

# About the Editors

**Marija Burinskienė**, Professor of Technological Science at Vilnius Gediminas Technical University, Lithuania. She has a PhD in Technological Science (1983). For more than 20 years, she was head of the Urban Engineering Department. For more than 10 years, she was a member of the Main Commission for civil engineers attestation. She is the author and co-author of 6 monographs and more than 120 research articles. She is a member of the editorial board of five international scientific research journals and the supervisor of 13 students' PhDs. She has supervised more than 130 national research and experimental development works and participated in more than 20 international projects, most of them in a position of the Lithuanian team. She has supervised the following: ARTS, ATLANTIC, VOJEGER, MOST 1997–2003; Save programme STEP by STEP 2003–2005; EU programme FRAMEWORK6- MAX and PROCEED in 2006–2009; 2011–2013: INTERREG Climate Change. Cultural Heritage & Energy. Efficient Monuments. Co2ol Brick; 2011–2014: TEMPUS: Development and improvement of automative and urban engineering studies in Serbia; DIAUSS; EESPON program 2010–2014; Swedish Institute program VISBY 2010–2016, INTERREG MARA (Mobility solutions for remoted areas) 2018–2021; COST-17125. Her research area is sustainable urban development, interaction between land use and planning systems and implementation of transportation infrastructure.

**Rasa Ušpalytė-Vitkūnienė**, Associate Professor. She holds a Master's degree in Civil Engineering (VGTU—2002) and a PhD in Technological Sciences (VGTU—2006). She is an Associate Professor at VGTU. Since the beginning of her doctoral study, she has been actively engaged in national projects and 5 international projects: MARA (Mobility solutions for remoted areas) Program INTEREG BJR, 2018–2020; ERASMUS MUNDUS (Preparing new join international master program in traffic safety EMPOWERS). Erasmus mundus, 2012–2014; BALTRIS (Improving Road Infrastructure Safety in the Baltic Sea Region). Program INTEREG, 2010–2012; PORTAL (Promotion of Results in Transport Research and Learning). Program FP5, 2001–2004; "Step by Step" (promoting cycling and public transport). Program "SAVE", 2003–2005; PROCEED (Principles of Successful High Quality Public Transport Operation and Development) Program FP6, 2006–2009. She is part of the board committee of the international conference "Environmental Engineering". The main field of interest is sustainable and safe transport planning in cities. She is the author and co-author of 25 journal publications, with a major part of them concerned with transport system planning and transport safety.

Article

# Comprehensive Traffic Calming as a Key Element of Sustainable Urban Mobility Plans—Impacts of a Neighbourhood Redesign in Ljutomer

**Mojca Balant [1,*] and Marjan Lep [2]**

[1] Urban Planning Institute of the Republic of Slovenia, 1000 Ljubljana, Slovenia
[2] The Faculty of Civil Engineering, Transportation Engineering and Architecture, University of Maribor, 2000 Maribor, Slovenia; marjan.lep@um.si
* Correspondence: mojcab@uirs.si

Received: 31 August 2020; Accepted: 28 September 2020; Published: 2 October 2020

**Abstract:** Negative impacts of motor vehicle traffic in cities are still increasing despite the objectives that sustainable mobility paradigm put forward almost three decades ago. Measures to reduce them still primarily focus on traffic safety improvements through vehicle speeds and flows reduction (traffic calming). Comprehensive traffic calming, a measure of sustainable urban mobility planning, targets the issue more comprehensively by also addressing changes in travel behaviour and quality of life. Literature covering the effects of measures addressing all listed aspects is scarce. In this paper, we present results of a longitudinal study of a comprehensive traffic calming redesign of a residential neighbourhood in Ljutomer in Slovenia. The following set of indicators was monitored: travel habits of neighbourhood residents, quality of living environment, acceptability of redesign, vehicle flows, speeds and classes, and traffic accidents. Motorized traffic counts, pilot interviews, postal and in-person surveys and public databases on traffic accidents were used to gather data before and after the redesign. All monitored indicators showed positive results. Around a third of residents claim to walk, cycle and socialize more than before the redesign while around two thirds state that the quality of life in the neighbourhood has improved. Vehicle speeds, flows and peak hour flows have notably decreased, and road safety has improved. The results show that the comprehensive traffic calming approach has a broad range of positive effects and contributes to achieving sustainable mobility. Its potential for a wider use in sustainable urban mobility planning practice is substantial.

**Keywords:** comprehensive traffic calming; active mobility; travel behaviour; quality of life; traffic safety; sustainable urban mobility planning; sustainable urban mobility plan; SUMP

## 1. Introduction

In recent years, sustainable urban mobility planning (SUM planning) has become increasingly established as a new approach for transport planning and mobility management in urban areas in a sustainable and comprehensive way. It follows the principles of the overarching sustainable mobility paradigm whose purpose is "to design cities of such quality and at a suitable scale that people would not need to have a car" [1] (p. 74). The paradigm also promotes the new transport hierarchy [1,2] from which the SUM planning takes its objectives of improving accessibility, quality of life and traffic safety, and increasing the use of sustainable travel modes. The latter are also objectives for urban mobility at the European level [3,4]. SUM planning approach is used for the preparation of Sustainable Urban Mobility Plans (SUMP), a strategic document that builds on existing planning practices while also considering integration, participation and evaluation principles [5]. The preparation and implementation of SUMPs is becoming a common practice in Europe and Slovenia in the last decade [5,6].

The development of a new paradigm of sustainable mobility is about thirty years old [7]. Its development has been stimulated by the constant increase in the volume of motorized traffic and, consequently, by its increasingly pronounced negative effects. While mobility has brought about positive economic and social effects, such as wealth, international collaboration, and exchange [8], there are also negative aspects including high proportion of urban land used by transport, urban sprawl, congestion, traffic noise, energy use and social and environmental problems [2,7–10]. Furthermore, major negative effects are mainly related to the private car [10]. Its intensive use has been proven to reduce the amount of physical activity, increase the possibility of traffic accidents, have a negative impact on health and the living environment and reduce the possibilities for social interaction [3,4,11–18].

Sustainable urban mobility planning addresses these challenges. Its main goal is to reduce the use of powered private vehicles. It focuses on sustainable travel modes, especially active mobility (walking and cycling), which is characterized by being the healthiest, least environmentally controversial, economically most rational and most socially equitable form of mobility [2,19–23]. These advantages make active mobility "the most favourable mode in terms of sustainability" [2] (p. 137) while it is also supported by other modern paradigms for creating green, healthy cities that are pleasant to live in [24,25].

Achieving a notable increase in active mobility for daily trips requires a significant improvement of the conditions for walking and cycling by establishing a system of safe, comfortable, direct and attractive infrastructure [23,26] and exclusive routes for pedestrians and cyclists [1]. Comprehensive traffic calming, a measure of SUM planning, is increasingly recognized as one of the more effective approaches. Its basic elements are larger set areas, most often in residential neighbourhoods, around schools and in city centres, where pedestrians and cyclists have priority. In addition to arrangements to reduce the speed and volume of motorized traffic aiming at improving traffic safety, the interventions also include the redesign of the public open space with the aim of improving the quality of living environment and changing travel habits into more sustainable ones. Several authors discuss the characteristics of this type of planning approach [9,18,27–33], and various older [34] and recent [35–40] examples of good practices from Northern and Western Europe are described. However, there is a lack of studies in the literature on the comprehensive quantified effects of (comprehensive) traffic calming on changing travel habits, traffic safety and quality of life [41–44].

The article presents the results of the comprehensive multi-year monitoring and evaluation of a pilot redesign of a residential neighbourhood in Ljutomer in Slovenia into an area with comprehensively calmed traffic. In 2014, 2017 and 2018 (before and after the redesign), the study systematically collected qualitative and quantitative data for the following indicators: travel habits of neighbourhood residents, quality of living environment, acceptability of redesign, vehicle flows, speeds and classes, and traffic accidents. The redesign of the neighbourhood was one of the measures of the first municipal SUMP [45], and the neighbourhood was in part also chosen because the municipality had already planned the renovation of underground municipal infrastructure and thus used the planned construction works for innovative improvement of traffic regulation. The redesign was carried out in line with the principles of SUM planning and followed the key steps for the preparation of the SUMP [5,46]. It is the first example of such a redesign and monitoring of effects in the country and the wider region. The importance of the study is even greater, as such a wide range of indicators is rarely measured for an individual measure [42].

### 1.1. Intensive Car Use and Health Issues

As pointed out above, intensive use of private car has been proven to reduce the amount of physical activity and has led to an increasingly sedentary life even when people essentially move around [47]. Consequently, this adds substantially to the general lack of physical activity, which has become one of the leading causes of death worldwide in recent years. As a key risk factor for non-communicable diseases, lack of physical activity claims more than a million premature deaths a year in Europe [17]. Comparatively, traffic accidents have a lower tax, but the numbers are still high.

In Europe, more than 25,000 people die on the roads every year, and around 200,000 suffer serious injuries [13]. The data show that conventional approaches to traffic calming in urban areas, which focus mainly on reducing driving speeds and improving traffic safety, are not effective enough. In Europe, almost 8000 pedestrians and cyclists still die in traffic accidents each year, and more than 60,000 are seriously injured. Nearly two thirds of accidents involving pedestrians and cyclists occur on roads within urban areas [13]. Thus, with their high population densities and high share of short-distance trips, the cities have the greatest potential to move towards sustainable travel modes [3] and to achieve the European Union's ambitious goals of increasing the share of active mobility on daily routes, reducing the number of traffic accidents, and improving the quality of life [11].

### 1.2. Pilot Neighbourhood Characteristics

The Juršovka residential neighbourhood is part of Ljutomer (3400 inhabitants) in Slovenia (Figure 1). It comprises four streets with one transit central axis and three access streets. The streets are connected by several segregated footpaths. The longest route is less than a kilometre and a half from one end of the neighbourhood to the other. The area lies on hilly terrain facing south. The neighbourhood consists of single-family houses with 119 households and 352 inhabitants. The majority of the population is aged between 18 and 65 (66.8%), 17.9% are under 18 and 15.3% are over 65 [48]. Prior to the redesign, the traffic regime restricted traffic with classic restrictions for settlements (permitted speeds of up to 50 km/h). Due to the small amount of traffic, no pavement was built in the neighbourhood, which bothered the residents, as individual drivers exceeded the speed limit and thus endangered pedestrians and cyclists. The area started experiencing parking on the road, which did not obstruct traffic, but was disruptive. Pedestrian connections were not maintained or lit.

**Figure 1.** The settlement of Ljutomer with the marked residential neighbourhood of Juršovka (Source of background picture: www.geoprostor.net).

With the comprehensive traffic calming, the neighbourhood was redesigned in terms of traffic arrangements and public open space. A different traffic regime was introduced, and new, quality public spaces were set up (Figure 2). The redesign was based on twelve typical elements: 30 km/h speed limit on the main (transit) street and 10 km/h on the side (access) streets; cycling without designated bicycle lanes due to low speeds; pavement on one side along the main street; renovation and lighting of the segregated footpaths; parking allowed only in marked places; setting up areas for socialising with benches and other urban equipment and playground equipment; plantings with trees, shrubs and perennials; traffic calming by converting intersections into mini roundabouts; traffic calming by a speed hump at the point of contact of the footpath with the road; traffic calming by narrowing the carriageway with combinations of plantings, planters, benches, playground equipment and parking spaces; traffic calming by narrowing entry points to the neighbourhood.

**Figure 2.** Three examples of the redesign elements showing the same locations before and after the redesign: (**a**) traffic calming by narrowing the carriageway with benches, playground equipment and plantings; (**b**) traffic calming by narrowing the carriageway with parking spaces and plantings and (**c**) traffic calming by converting intersections into mini roundabouts.

## 2. Materials and Methods

The effects of comprehensive traffic calming in the Juršovka neighbourhood in Ljutomer were analysed using empirical research, namely through surveys, pilot interviews, time series analysis and analysis of data in public databases. Data were collected before and after the redesign of the neighbourhood, which took place in 2016 and was completed in July 2016.

The survey and pilot interviews were used to identify changes in the travel habits of neighbourhood residents. The survey was also used to determine the acceptability of the redesign of the neighbourhood into an area of comprehensively calmed traffic and changes in the perceived quality of the living environment. We conducted surveys and interviews in 2014 and 2017. The pilot interviews were conducted on 22 May 2014 as part of the preparation of a conceptual project neighbourhood redesign [48]. We interviewed 10 households (8%), which were selected on the basis of demographic and spatial analysis. The selection ensured an even representation of age groups (0–18, 18–65 and over 65 years of age) and the distribution of interviewees throughout the neighbourhood. The first survey was conducted in June 2014, following a workshop on the conceptual design of the neighbourhood redesign and as part of the preparation of a conceptual project for neighbourhood redesign [48]. It was sent to all households and was completed by 30 households (25%). The survey was simple and short, its key part was a questionnaire about the support of the proposed typical redesign elements. Respondents rated individual elements using a three-point scale. The second survey was conducted in June and July 2017 as part of the activities of the European project Civitas Prosperity [49]. The survey was delivered in person to all 119 households in the neighbourhood and was completed by 85 households (71%). The survey was longer and more complex; we collected data on changes in travel habits and the quality of the living environment, as well as responses to typical redesign elements and the redesigned neighbourhood as a whole. Respondents rated the answers using a five-point scale. We used time series analysis to determine changes in vehicle flows, speeds and classes in three-time sections. The situation before the redesign was recorded on 17 June 2014 [50], and the situation one and two years after the redesign was recorded on 6 July 2017 [51] and 26 September 2018 [52]. Data from public databases were used to determine the number and consequences of traffic accidents in the neighbourhood and in the settlement of Ljutomer. We obtained them from the statistical files of the National Police containing data on all traffic accidents in Slovenia for the period 2000–2019 [53].

### 2.1. *Travel Habits of Neighbourhood Residents*

Data on the travel habits of neighbourhood residents were collected through pilot interviews (2014) and a household survey (2017). We were mainly interested in the use of active mobility (walking and cycling) on daily routes. The interviewees specified for the whole household the use of travel modes for different daily routes (to work, running errands, to school and to/for recreation) and the reasons for not using walking and cycling on daily routes. Respondents to the survey specified for three age groups within an individual household (0–18, 18–65 and over 65 years of age), using a five-point response scale, the more frequent use of walking and cycling on daily routes after the redesign and the contribution of individual elements of the redesign to the facilitation of walking and cycling.

### 2.2. *Quality of Living Environment*

Data on the quality of the living environment in the neighbourhood after the redesign were collected through a household survey (2017). Using a five-point response scale, respondents rated the improvement of the overall quality of life in the neighbourhood, the frequency of staying outside and socialising with neighbours after the redesign and the contribution of each redesign element to improving the quality of living environment.

### 2.3. *Acceptability of the Redesign*

Data on the acceptability of the redesign of the neighbourhood into an area of comprehensively calmed traffic were collected through household surveys (2014 and 2017). In the first survey, respondents rated nine proposed typical redesign elements using a three-point response scale. The part of the second survey relating to the acceptability of the redesign encompassed twelve redesign elements that were actually used, including all elements from the first survey. Respondents rated their support for each typical redesign element using a five-point response scale and, according to the characteristics of each element, an appropriate set of the following categories: understandability, compliance with the regime, use, impact on traffic calming. They could also list the parts of the redesign they liked most and least.

### 2.4. *Vehicle Flows and Speeds*

All three-time sections of motorized traffic measurements are data obtained at the same four locations within the neighbourhood. Three measuring locations were on the main (transit) street through the neighbourhood, one of which was at the entrance point to the neighbourhood, and one measuring location was on a side (access) street. Data were collected for the needs of the research work of the group for sustainable mobility from the Urban Planning Institute of the Republic of Slovenia, Ljubljana, Slovenia. The measurements were carried out by the Centre for Mobility Research from the Faculty of Civil Engineering, Transportation Engineering and Architecture of the University of Maribor, Maribor, Slovenia. Measurements were performed on a working day with automatic pneumatic traffic counters. Daily (24 h) two-way vehicle flows, speeds and classes were measured at all four measuring locations.

### 2.5. Traffic Safety

We were interested in traffic accidents in the area of the Juršovka neighbourhood and comparatively in the entire settlement of Ljutomer. For Ljutomer, we obtained data on the number of accidents, number of participants and serious consequences (severe injury or death, type of participant). We obtained some more detailed data for the neighbourhood (all types of consequences, including material damage and minor injury, cause and type of accident, type of participant, location of the accident (street)).

## 3. Results

### 3.1. Travel Habits of Neighbourhood Residents

#### 3.1.1. Travel Habits of Neighbourhood Residents Before Redesign

Most residents use a car for daily trips to work and school and for running errands (80%); walking and cycling are rarely chosen (15% and 4%, respectively) and are used most often for running errands (23% on foot, 6% by bicycle). For most people, walking and cycling are a form of afternoon recreation (56% of them walk and 32% cycle recreationally or to get to recreation).

The most cited reasons for the infrequent use of walking and cycling for daily trips were lack of time (e.g., they run errands by car so they have more time for afternoon recreation in nature), the dangerous route through the settlement, especially for children (no pavement, excessive speeds, dangerous intersections, obstructed views of the road), long distances, sloped terrain and the fact that it is not common to take daily trips on foot or by bicycle.

#### 3.1.2. Travel Habits of Neighbourhood Residents after Redesign

The use of walking and cycling for daily routes was significantly higher one year after the redesign of the neighbourhood (Table 1). 37% of the population walk more often (18% a lot and 19% a little more often), 27% cycle more often (15% a lot and 12% a little more often). After the redesign, 2% walk less often and 11% cycle less often, while 61% walk and cycle the same as before the redesign. The frequency of walking and cycling increased the most among children and youth (0–18 years); 40% walk more often (10% a lot and 30% a little more often), 36% cycle more often (18% a lot and 18% a little more often). They are followed by adults (18–65 years); 39% walk more often (21% a lot and 18% a little more often), and 31% cycle more often (16% a lot and 15% a little more often). The smallest increase in the use of walking and cycling was reported by the elderly (over 65 years); 30% of them walk more often (15% a lot and 15% a little more often), and only 8% cycle more often (8% a lot and 0% a little more often).

**Table 1.** Frequency of walking and cycling after redesign of the neighbourhood by individual age groups.

| Travel Mode | Age Group | YES, Very Often | YES, Quite Often | YES, a Little More Often | Same as Before | NO, Less Often |
|---|---|---|---|---|---|---|
| Walking | All | 13% | 4% | 19% | 61% | 2% |
| | Children and youth (up to age 18) | 10% | 0% | 30% | 60% | 0% |
| | Adults (ages 18–65) | 15% | 6% | 18% | 59% | 3% |
| | Elderly (over 65) | 12% | 4% | 15% | 69% | 0% |
| Cycling | All | 11% | 4% | 12% | 61% | 11% |
| | Children and youth (up to age 18) | 14% | 5% | 18% | 55% | 9% |
| | Adults (ages 18–65) | 12% | 4% | 15% | 60% | 9% |
| | Elderly (over 65) | 4% | 4% | 0% | 72% | 20% |

The majority of the population (58%) believe that the redesign of the neighbourhood into an area of comprehensively calmed traffic encourages more frequent use of walking and/or cycling (25% disagree, 18% are undecided). According to residents, the following elements have the most positive impact: pavement along the main street through the neighbourhood (79%); renovation of segregated footpaths

(77%); 30 km/h speed limit on the main street (62%); cycling without designated bicycle lanes due to low car speeds (62%) and plantings with trees, shrubs and perennials (58%).

### 3.2. Quality of Living Environment

The majority of the population (63%) believe that the overall quality of life in the neighbourhood has significantly improved after the redesign, 19% believe that it has not improved, and 19% are undecided (Table 2). Elements of redesign that, according to residents, contributed the most to improving the overall quality of life in the neighbourhood are the pavement along the main street through the neighbourhood (87%); renovation of footpaths (84%); plantings with trees, shrubs and perennials (69%); traffic calming by speed humps (68%) and the 30 km/h speed limit on the main street (66%). Even for most of the remaining elements of the redesign, more than half of the population was of the opinion that they had contributed to improving the overall quality of life in the neighbourhood. The worst rated elements were the conversion of intersections into mini roundabouts (36% in favour and 39% against) and traffic calming by narrowing the carriageway on both entry points to the neighbourhood (49% in favour and 28% against).

**Table 2.** Change in overall quality of life after neighbourhood redesign.

| | I Strongly Agree | Agree | Neither Agree nor Disagree | Disagree | Strongly Disagree |
|---|---|---|---|---|---|
| The redesign has improved the overall quality of life in the neighbourhood | 25% | 38% | 19% | 10% | 9% |

The frequency of staying outside and socialising with neighbours was substantially higher one year after the redesign of the neighbourhood (Table 3). Of the population, 28% spend more time outside and socialise with neighbours (14% much and 14% a little more often), 3% less often, and the majority (69%) the same as before the redesign. Adults (18–65 years) changed their habits the most. After redesign, just under a third (32%) of them spend more time outside and socialise with neighbours. The share of the elderly (over 65 years) is 26%, and among children and youth (0–18 years), it is 19%. The majority maintained their habits (64% of adults, 74% of the elderly and 81% of children and youth), and only 4% of adults reported less frequent stays outside and socialising with neighbours after the redesign.

**Table 3.** Changes in the frequency of staying outside and socialising with neighbours after neighbourhood redesign by individual age groups.

| | | YES, Very Often | YES, Quite Often | YES, Somewhat More Often | Same as Before | NO, Less Often |
|---|---|---|---|---|---|---|
| Staying outside and socialising with neighbours | All | 8% | 7% | 14% | 69% | 3% |
| | Children and youth (up to age 18) | 5% | 5% | 10% | 81% | 0% |
| | Adults (ages 18–65) | 9% | 6% | 17% | 64% | 4% |
| | Elderly (over 65) | 7% | 11% | 7% | 74% | 0% |

### 3.3. Acceptability of the Redesign

#### 3.3.1. Acceptability of the Redesign, before Redesign

On average, almost four fifths of the population (78%) expressed support for the proposed typical redesign elements, 8% of the population expressed partial support and 14% opposed the changes. Four of the nine redesign elements received very high levels of support with negligible opposition, while the other half received the support of more than half of the population, but also opposition from at least a fifth.

A very high level of support was given to the pavement along the main street through the neighbourhood (100% in favour), traffic calming at 30 km/h on the main street (90% in favour and 0% against), cycling without designated bicycle lanes due to low car speeds (90% in favour and 7% against) and side streets without raised pavements (83% in favour and 7% against). The least support was given to traffic calming by narrowing of the carriageway with plantings, benches, playground equipment or parking spaces (57% in favour and 32% against) and traffic calming at 10 km/h on side streets (65% in favour and 28% against); these were followed by parking only in marked places in the neighbourhood (63% in favour and 20% against) and converting intersections into mini roundabouts (72% in favour and 21% against).

### 3.3.2. Acceptability of the Redesign, after Redesign

Support for the typical redesign elements remained high. On average, three quarters of the population (75%) agree with the new neighbourhood layout, 15% oppose it, and 10% are undecided. Of the twelve typical redesign elements used, five received very high levels of support with little opposition, six received support from more than half of the population but in some cases received increased opposition, and one element was rated very negatively (Figure 3).

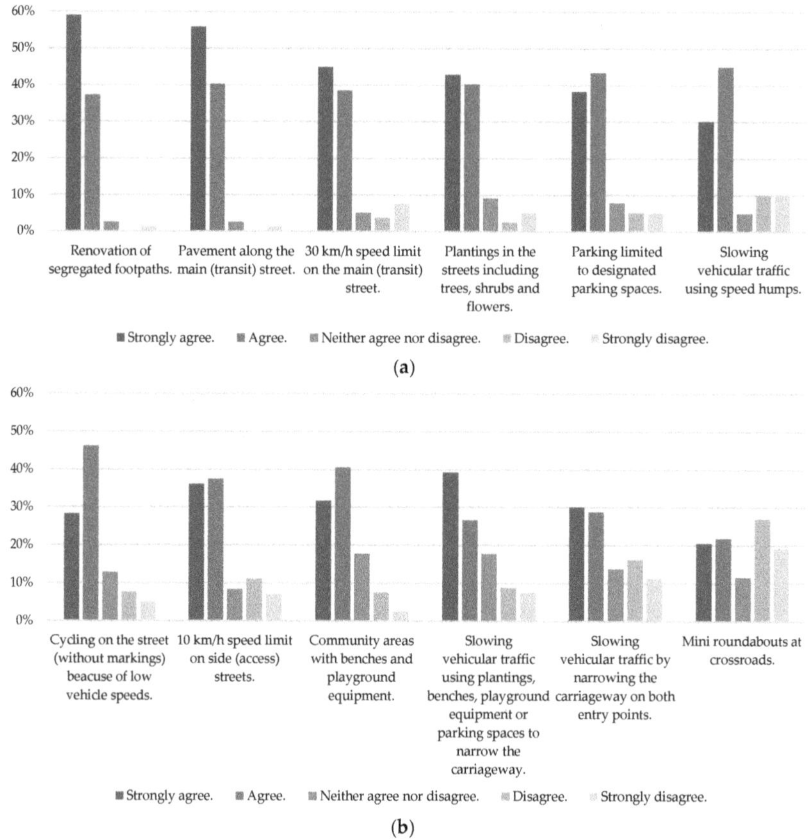


**Figure 3.** Rating of acceptability among the population for all twelve typical elements of redesign from the element with the most support to the one with the least support; (**a**) elements with more support and (**b**) elements with less support.

Very high levels of support were given to the building of the pavement along the main street through the neighbourhood (93% in favour and 1% against), the renovation of the footpaths in the neighbourhood (93% in favour and 1% against), traffic calming at 30 km/h on the main street (83% in favour and 12% against), plantings with trees, shrubs and perennials (83% in favour and 8% against) and parking only in marked places in the neighbourhood (82% in favour and 11% against). The least support was given to the conversion of intersections into mini roundabouts (42% in favour and 46% against). The remaining elements of the redesign were supported by 59% to 75% of the population, while 13% to 28% of the population opposed them. From more to less acceptable are traffic calming by speed humps (75% in favour and 20% against), cycling without designated bicycle lanes due to low car speeds (74% in favour and 13% against), traffic calming at 10 km/h on side streets (74% in favour and 18% against), community areas with benches and arrangements for playing (72% in favour and 10% against), traffic calming with plantings, benches, playground equipment or parking spaces used to narrow the carriageway (66% in favour and 16% against) and traffic calming with narrowing at both entry points to the neighbourhood (59% in favour and 28% against).

Understandability was rated for limiting driving speed and mini roundabouts. Mini roundabouts proved to be problematic, with 43% of the population saying they were incomprehensible (comprehensible for 41%, 16% undecided). They were also by far the most frequently mentioned among the least likeable parts of the redesign. The speed limit is understandable to most residents (79% for 30 km/h and 70% for 10 km/h).

Adherence to the regime was rated for speed limit and parking. Residents claim that 96% adhere to the 30 km/h limit (1% does not, 3% are undecided) and 72% adhere to the 10 km/h limit (15% do not, 13% are undecided). According to the residents, 57% of the population adhere to the parking regime by only parking in marked places in the neighbourhood (30% do not, 13% are undecided). As the least likeable parts of the redesign, the residents have repeatedly stated that excessive speeds and illegal parking are still a problem and that it is necessary to introduce consistent speed control and more frequent traffic warden control.

The use was rated for parking spaces (residents and visitors), socializing and play areas (adults and children), pavement along the main street and segregated footpaths. Residents mostly use footpaths (96%) and the pavement (95%). Socializing and play areas are used by 70% of children and 50% of adults. Parking spaces are used by more than half of the population (54%) and their visitors (55%).

The impact of typical redesign elements on traffic calming was the category that received the lowest values. On average, only a little more than half of the population believes that redesigns have had an impact on traffic calming (59%), 26% do not, and 15% are undecided. According to the residents, the most effective element were speed humps (68% in favour and 22% against). This is followed by narrowing at the entry points to the neighbourhood (64% in favour and 27% against); narrowing of the carriageway with plantings, benches, playground equipment or parking spaces (62% in favour and 25% against) and the 30 km/h speed limit on the main street (59% in favour and 27% against). In the context of the impact on traffic calming, the pavement along the main street (53% in favour and 17% against) and building mini roundabouts (48% in favour and 36% against) were rated the lowest. The problematic nature of the latter was already detected in terms of acceptability and understandability.

The acceptability of the redesign of the neighbourhood into an area of comprehensively calmed traffic was also measured with questions about its most and least likeable part. Just under a quarter of the population likes the redesign in its entirety (24%), while the other most frequently mentioned likeable elements are areas for socialising and playing (14%), pavement for its safety (12%), segregated footpaths (10%) and side streets due to both calmed traffic and areas for socialising and playing (10%). On the other hand, almost a third of the population likes mini roundabouts the least (29%), followed by the narrowing of the carriageway (8%), paving stones on the carriageway (5%) and speed humps (5%). These elements disturb the residents mainly because they hinder smooth driving. Among the criticisms were also the following more general comments: poor quality of construction works, poor maintenance of plantings and other arrangements, the need for consistent speed control to prevent speeding and for more frequent traffic warden control to prevent illegal parking.

### 3.4. Vehicle Flows and Speeds

Measurements of motorized traffic were carried out at four measuring locations in Juršovka residential neighbourhood in Ljutomer in Slovenia (Figure 4). Three measuring locations were on the main (transit) street through the neighbourhood: at 22 Kidričeva Street (location 1; micro-location after redesign: longer straight section, near the narrowing of the carriageway with parking spaces, an area with benches and a speed hump), 17 Jurčičeva Street (location 2; micro-location after redesign: longer straight section, between a speed hump and a narrowing of the carriageway with a parking space) and 1 Aškerčeva Street (location 3; entry point to the neighbourhood, micro-location after redesign: short straight section between two mini roundabouts). One measuring location was on a side (access) street at 11 Aškerčeva Street (location 4; micro-location after redesign: longer straight section near the narrowing of the carriageway with a parking space).

**Figure 4.** Juršovka residential neighbourhood in Ljutomer in Slovenia with marked measuring locations. Location 1: 22 Kidričeva Street, location 2: 17 Jurčičeva Street, location 3: 1 Aškerčeva Street and location 4: 11 Aškerčeva Street (source of background picture: www.geoprostor.net).

#### 3.4.1. Vehicle Flows and Classes before and after Redesign

Data on the number of vehicles travelling towards the centre and away from the centre, for both directions together and on the peak hour flows for all four measuring locations and all three-time sections are shown in Table 4. The amount of traffic decreased throughout the neighbourhood after the redesign. The lowest number of vehicles was recorded in 2017, which may also be due to seasonal changes (measurements were carried out in July, when people already begin taking their leaves for summer holidays), but the quantities in 2018, when measurements were carried out in September, are only slightly higher than in 2017 and significantly lower than in 2014. As expected, the entry point to the neighbourhood is the busiest, and the access street is the least crowded. Vehicle flows proportionally decreased the most at the entry point and the least on the access street, which is not surprising since the latter is a characteristic route destination.

**Table 4.** Vehicle flows and peak hour flows during measurements before (2014) and after (2017 and 2018) redesign showing the change for the period 2014–2018 for all four measuring locations.

| Measuring Location | | Number of Vehicles in 24 h | | | |
|---|---|---|---|---|---|
| | | 17 June 2014 | 06 July 2017 | 26 September 2018 | Change 2014–2018 |
| Location 1; Kidričeva Street (main street, transit) | Direction toward centre | 163 | 110 | 118 | −45 (−28%) |
| | Direction away from centre | 165 | 100 | 129 | −36 (−22%) |
| | Both directions | 328 | 210 | 247 | −81 (−25%) |
| | Peak hour flow | 39 | 19 | 28 | −11 (−11%) |
| Location 2; Jurčičeva Street (main street, transit) | Direction toward centre | 287 | 115 | 148 | −139 (−48%) |
| | Direction away from centre | 277 | 124 | 169 | −108 (−39%) |
| | Both directions | 564 | 239 | 317 | −247 (−44%) |
| | Peak hour flow | 55 | 23 | 32 | −23 (−23%) |
| Location 3; Aškerčeva Street (main street, transit, entry point) | Direction toward centre | 444 | 282 | 291 | −153 (−34%) |
| | Direction away from centre | 431 | 325 | 307 | −124 (−29%) |
| | Both directions | 875 | 607 | 598 | −277 (−32%) |
| | Peak hour flow | 95 | 65 | 67 | −28 (-29%) |
| Location 4; Aškerčeva Street (side street, access) | Direction toward centre | 78 | 77 | 73 | −5 (−6%) |
| | Direction away from centre | 75 | 65 | 77 | 2 (+3%) |
| | Both directions | 153 | 142 | 150 | −3 (−2%) |
| | Peak hour flow | 18 | 14 | 17 | −1 (−6%) |

From 2014 to 2018, the number of vehicles per working day on transit streets decreased by an average of 33%. In 2018, there were 277 fewer vehicles recorded at the entry point to the neighbourhood (−32% compared to 2014); on the two transit streets, there were 247 and 81 fewer vehicles (−44% and −25%, respectively, compared to 2014), and on the access street, there were 3 fewer vehicles (−2% compared to 2014). The average peak hour flow in the neighbourhood also decreased by just under a third (−30%) from an average of 52 vehicles per hour in 2014 to an average of 36 in 2018. On transit streets, the average peak hour flow was reduced from 63 vehicles per hour in 2014 to 42 in 2018 (−33%); the peak is between 15:00 and 16:00. On the access street, the peak hour flow remained practically the same (18 in 2014 and 17 in 2018) (−6%), with the peak between 14:00 and 15:00.

The traffic counters used classify vehicles into 12 classes. At the time of the measurements, three different vehicle classes were recorded in the neighbourhood. The shares of each class are given below for the average of all measuring locations for each year of measurements. The majority of vehicles in the neighbourhood are passenger cars (vehicle class 2), namely, 99% in 2014 and 2017 and 98% in 2018. The remaining 1% and 2%, respectively, are passenger cars with trailers and two-axle lorries or buses (vehicle classes 3 and 4); most of these vehicles were recorded on transit streets.

### 3.4.2. Vehicle Speeds before and after Redesign

Data on the proportions of passenger cars (vehicle class 2) according to the measured driving speed for all four measuring locations and all three-time sections are shown in Table 5. In general, the speed of passenger cars has significantly decreased throughout the neighbourhood after the redesign, although they have not yet reached the permitted speed limits, especially not on side streets with a speed limit of 10 km/h. The smallest changes in vehicle speeds were recorded at the entry point to the neighbourhood, where in all time sections most vehicles drove at speeds up to 30 km/h, probably due to the micro-location between two intersections or mini roundabouts (depending on the situation before and after redesign).

Only a small number of passenger cars drive less than 10 km/h in the neighbourhood (2% in 2014 and 1% in 2018), but the share of vehicles driving slower than 30 km/h has increased on average by 20% (from 38% in 2014 to 58% in 2018), and the share of those who drive faster (30–50 km/h) decreased proportionally (from 59% in 2014 to 42% in 2018). Speeds above 50 km/h are practically non-existent (3% in 2014 and 0.2% in 2018). The median speed on transit streets decreased on average by 6.0 km/h and on the access street by 2.8 km/h.

**Table 5.** Driving speeds of passenger cars during measurements before (2014) and after (2017 and 2018) redesign showing the change for the period 2014–2018 for all four measuring locations.

| Measuring Location | | Share of Passenger Cars in Relation to Measured Driving Speed | | | |
|---|---|---|---|---|---|
| | | Up to 10 km/h | Up to 30 km/h | 30–50 km/h | Over 50 km/h |
| Location 1; Kidričeva Street (main street, transit) | 17 June 2014 | 0% | 16% | 74% | 10% |
| | 06 July 2017 | 1% | 41% | 55% | 3% |
| | 26 September 2018 | 0% | 36% | 62% | 1% |
| | Change 2014–2018 | 0% | +20% | −11% | −9% |
| Location 2; Jurčičeva Street (main street, transit) | 17 June 2014 | 0% | 17% | 79% | 4% |
| | 06 July 2017 | 0% | 70% | 29% | 0% |
| | 26 September 2018 | 1% | 59% | 41% | 0% |
| | Change 2014–2018 | 1% | +42% | −39% | −4% |
| Location 3; Aškerčeva Street (main street, transit, entry point) | 17 June 2014 | 1% | 98% | 2% | 0% |
| | 06 July 2017 | 1% | 85% | 11% | 4% |
| | 26 September 2018 | 1% | 100% | 0% | 0% |
| | Change 2014–2018 | 0% | +2% | −2% | 0% |
| Location 4; Aškerčeva Street (side street, access) | 17 June 2014 | 2% | 32% | 67% | 1% |
| | 06 July 2017 | 6% | 61% | 38% | 1% |
| | 26 September 2018 | 1% | 50% | 50% | 0% |
| | Change 2014–2018 | −1% | +18% | −17% | −1% |

To show the change in passenger cars' speed through the neighbourhood, the most representative measuring location is 22 Kidričeva Street (Figure 5), which is located on a long straight section without major intersections and after redesign is close to two typical elements (narrowing of the carriageway by parking spaces and bench areas along the carriageway and a speed hump). Prior to the redesign, most passenger cars drove through the neighbourhood at speeds between 30 and 50 km/h ($V_{85}$ in 2014 was 48.25 km/h), and after the redesign, most drove at speeds between 20 and 40 km/h ($V_{85}$ in 2018 was 41.28 km/h). A comparison of the cumulative shares of passenger cars for individual driving speeds for the period before (2014) and after (2017 and 2018) redesign for all four measuring locations is shown in Figure 6.

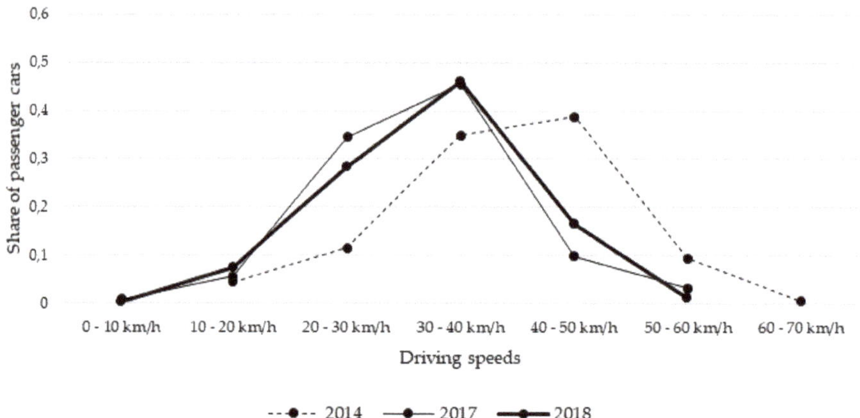


**Figure 5.** Driving speeds of passenger cars at the 22 Kidričeva Street measuring location during the measurements before (2014) and after (2017 and 2018) redesign.

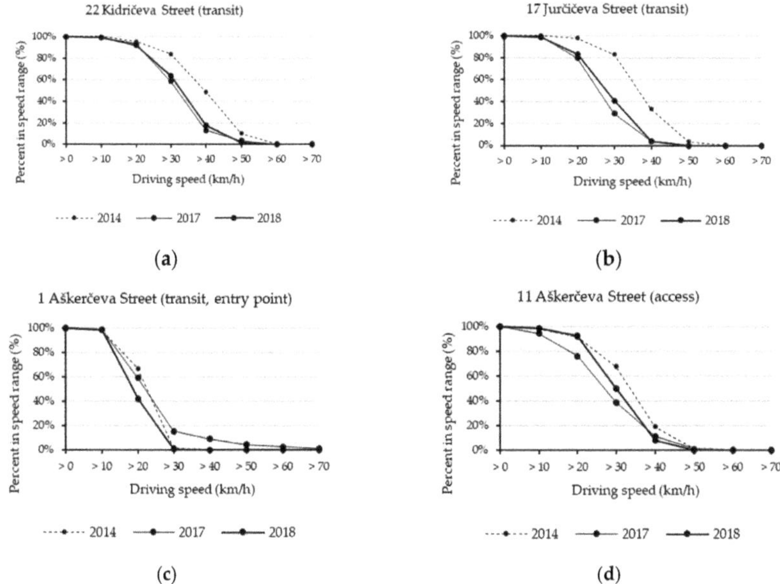


**Figure 6.** Comparison of cumulative shares of passenger cars for individual driving speeds for the period before (2014) and after (2017 and 2018) redesign for all four measuring locations; (**a**) location 1: 22 Kidričeva Street, (**b**) location 2: 17 Jurčičeva Street, (**c**) location 3: 1 Aškerčeva Street and (**d**) location 4: 11 Aškerčeva Street.

There are few larger vehicles of vehicle classes 3 and 4 in the neighbourhood (1.1% (23) in 2014, 0.8% (9) in 2017 and 1.5% (20) in 2018). A comparison of 2014 and 2018, with practically the same number of larger vehicles, showed a reduction in driving speed in the neighbourhood. In 2018, 75% of larger vehicles drove slower than 30 km/h (52% in 2014). No larger vehicle drove faster than 50 km/h either before or after the redesign.

### 3.5. Traffic Safety

In the period 2000–2019, there were 802 traffic accidents with 1310 participants in the settlement of Ljutomer, 17 participants had serious injuries (including one pedestrian and four cyclists); there were no fatalities. Regarding traffic accidents, the Juršovka neighbourhood is not a problematic area. In the same period, there were only 12 traffic accidents (1.5% of the settlement total) with 21 participants (1.6% of the settlement total), all of which were without serious injuries or death, and only one participant had minor injuries. A comparison of data for the settlement of Ljutomer and the Juršovka neighbourhood before and after the redesign is given in Table 6 below.

No pedestrians or cyclists were involved in the accidents in the neighbourhood. The most common causes were unadjusted speed (25%) and incorrect side or direction of travel (25%), which in most cases resulted in a collision into an object (45%) or a collision of vehicles (27%). The majority of accidents (11, or 92%) occurred on transit streets (6 on Kidričeva Street, 4 on Mestni breg next to the entry point to the neighbourhood at Aškerčeva street, and one on Jurčičeva Street). One accident occurred on the access street (Aškerčeva Street). Most accidents in the neighbourhood occurred before the redevelopment (11, or 92%, between 2000 and 2010), and after the redevelopment only one (in 2018 on Kidričeva Street, a vehicle collided with a building due to the wrong side or direction of travel).

**Table 6.** Traffic accidents in the settlement of Ljutomer and the Juršovka neighbourhood in the period before (2000–2016) and after (2016–2019) redesign.

| Period and Monitored Categories | | Traffic Accidents—Locations, Number and Shares of Monitored Categories | | |
|---|---|---|---|---|
| | | Ljutomer Settlement | Juršovka Neighbourhood | Share in Relation to the Whole Settlement |
| Before redesign (2000–2016) | Number of accidents | 684 | 11 | 1.6% |
| | Number of participants | 1158 | 20 | 1.7% |
| | Death | 0 | 0 | - |
| | Serious injury | 16 | 0 | - |
| | Seriously injured participants | | | |
| | - Pedestrian | 1 | 0 | - |
| | - Bicyclist | 4 | 0 | - |
| | - Motorcyclist | 2 | 0 | - |
| | - Driver of a passenger car | 9 | 0 | - |
| | - Driver of a freight vehicle | 1 | 0 | - |
| After redesign (2016–2019) | Number of accidents | 118 | 1 | 0.8% |
| | Number of participants | 152 | 1 | 0.7% |
| | Death | 0 | 0 | - |
| | Serious injury | 1 | 0 | - |
| | Seriously injured participants | | | |
| | - Motorcyclist | 1 | 0 | - |

## 4. Discussion

Almost thirty years after the concept of sustainable mobility appeared on the international agenda, we are still far away from achieving a sustainable mobility system [7], and predominantly dependent on private car for mobility [10,54]. To avert this trend, we must focus on people and plan for them [1,55]. Several authors agree that neighbourhoods are among the most important places to do that [1,23,56]. The redesign of the Juršovka neighbourhood in Ljutomer is a step in this direction.

Comprehensive traffic calming measures were used for the neighbourhood redesign, following the principles of sustainable mobility paradigm [1] and SUM planning. The results of monitoring and rating by the residents show positive effects towards achieving the key objectives, namely, improving accessibility, quality of life and traffic safety, and increasing the use of sustainable travel modes [1–4]. Positive effects for a similar broader set of indicators are also reported in a study from the United Kingdom [57], which monitored and evaluated nine pilot areas for the conversion of residential neighbourhoods into calm residential zones (so-called Home zones), and in a report from Vitoria-Gasteiz in Spain [58], which examined the setup of a pilot traffic-friendly area (so-called Superblock). These two documents are a rare example of monitoring and evaluating a wider range of comprehensive traffic calming effects as opposed to a large number of studies focusing on traffic safety and vehicle flows and speeds reduction [23,59–65].

All three studies report an increase in the use of active mobility and an improvement of the quality of living environment. In Ljutomer, after the redesign, a little over a third of the population (37%) walk more often, and just under a third (27%) cycle more often. Youth under the age of 18 have changed their habits the most, followed by adults and the elderly. Just under two thirds of the population (63%) believe that the quality of living in the neighbourhood has improved significantly. Most of them show strong support for the redesign elements (75%), and just under a quarter like the redesign in its entirety (24%). The higher quality of the living environment is also confirmed by the fact that after the redesign 28% of the population spends more time outside and socialising with neighbours, and that 70% of children and 50% of adults use community and play areas. According to residents, the key elements of the redesign that contributed most to the above effects are the pavement along the main street through the neighbourhood; renovation of segregated footpaths; plantings with trees, shrubs and perennials and the 30 km/h speed limit on the main street. Residents agree the least with the mini roundabouts and traffic calming by narrowing at both entry points to the neighbourhood.

The study from United Kingdom [57] states that 10% of the population reported more frequent cycling, 44% thought that walking after the redesign was more pleasant while 12% spent more time outside. Of the population, 64% supported the redesign, 73% of them thought that their living environment was now more attractive. By far the most desirable element proved to be plantings with trees, shrubs and perennials. The pilot redesign in Vitoria-Gasteiz [58] received a score of 7.4 on a ten-point scale of acceptance among the population, and the redesign greatly affected the change in travel habits. The elimination of transit traffic from the area helped to reduce motorized traffic to less than 20%, while the number of pedestrians increased by 57% and the number of cyclists by 9%.

Residents rated highly the elements of the redevelopment, such as trees, exclusive routes for walking and cycling, quality public spaces, urban equipment, and public green spaces. These elements are an important added value that comprehensive traffic calming adds to the more technical elements of traffic calming. The results support paradigms of creating streets as spaces for people and not for cars [1,31,55,66], and the same elements are recognized in the literature as important factors in providing a supportive environment for the more frequent use of active mobility and increasing health benefits [20]. Synergies with other modern paradigms of establishing green, healthy cities that are pleasant to live in [24,25] also show the potential of linking sustainable urban mobility planning with the planning of public green spaces.

Redesigns of the studied areas also had a positive effect on reducing vehicle flows and speeds. In Ljutomer, the number of vehicles per working day decreased on transit streets on average by 33% and on the access street by 2%. The average peak hour flow decreased by almost a third (−30%) as well. Driving speeds have also decreased, although the target speeds of 30 km/h on transit and 10 km/h on access streets have not (yet) been reached. The $V_{85}$ at the representative measuring location decreased by 6.97 km/h to 41.3 km/h. The median speed on transit streets decreased on average by 6.0 km/h, and on the access street by 2.8 km/h. The share of vehicles driving slower than 30 km/h increased on average by 20% (to 58%), while the share of those driving between 30 and 50 km/h decreased proportionally (from 59% to 42%). The speed practically does not reach above 50 km/h (0.2%), but almost no one drives below 10 km/h (1%) either. The residents still perceive speeds as fairly high, as only a little over a half of them believe that the redesign has had an impact on traffic calming. Perception may be based on rare speeding vehicles, which residents tend to remember more than the majority of vehicles driving slowly [57]. On the other hand, the residents are not self-critical enough, with 96% claiming to comply with the 30 km/h limit and 72% claiming to comply with the 10 km/h limit.

The aforementioned study from the United Kingdom [57] measured a smaller reduction in the volume of traffic on transit streets (by 25%), while traffic calming was more effective. The $V_{85}$ decreased on average by 9.7 km/h to less than 30 km/h, and the median speed decreased on average by 8.0 km/h, but the starting speeds were lower (58% of vehicles drove slower than 30 km/h before the redesign and 88% after). Interestingly, only 20% of the population there reported driving slower after the redesign. The average speed of motorized traffic in Vitoria-Gasteiz [58] decreased by 2.2 km/h, but the speeds there were already lower as well before the redesign (below 30 km/h). Other studies report speed drops ranging from up to 6 km/h in Denmark [67] to up to 11 km/h in the United States and up to 18 km/h in the United Kingdom [60].

It is likely that the reduction of driving speeds in the neighbourhood results from both the reduction in vehicle flows (i.e., residents driving slower than vehicles in transit) as well as the effect of redesign elements. Measuring locations were located nearby different redesign elements with consequently different effects on reducing driving speeds. Location 2 (17 Jurčičeva Street) showed to be most successful in speed reduction. There, share of vehicles driving slower than 30 km/h increased by 53% in short-term and by 42% in mid-term. This measuring location was directly between two redesign elements (a speed hump and a narrowing of the carriageway with a parking space). We suppose this is the reason for more rigorous reduction in driving speeds since other two locations on similar sections of streets (location 1 at 22 Kidričeva Street and location 4 at 11 Aškerčeva Street) were located on one side of one or a combination of several redesign elements. There, smaller speed reductions were measured (share of vehicles driving slower than 30 km/h increased by 25% and 29% in short-term and 20% and 18% in mid-term, respectively). While very few vehicles drove faster than 30 km/h before the redesign at location 3 (1 Aškerčeva Street), this is the only location where all passenger cars (100%) drove slower than the target speed of 30 km/h in mid-term (85% in short-term). Its micro-location after the redesign is between two redesign elements (in this case two mini roundabouts). Interestingly, the latter present the most negatively accepted and least understood redesign element according to residents' responses. Therefore, it is likely that in the short-term some residents were driving directly over the mini roundabouts and consequently reached a higher speed at the measuring location while in the mid-term they started to use the mini roundabouts correctly followed by slower driving in-between them. All locations show trend inversion in speed reduction when comparing short- and mid-term measurements. It is most likely that this occurred due to residents getting familiar with the redesign elements. It would be interesting to measure driving speeds and implement a survey on the perception of the redesign elements to gather information and data on long-term effects of the redesign.

The last indicator is traffic safety. From the point of view of traffic accidents, the Juršovka neighbourhood in Ljutomer is not a problematic area, but the fact that 92% of accidents occurred before the redesign still speaks in favour of the redesign. Between 2000 and 2019, there were 12 traffic accidents in the neighbourhood (1.5% of the settlement total). All were without serious injury or death, and no pedestrians or cyclists were involved. After the redesign, only one accident occurred (in 2018 on a transit street, a vehicle collided into a building). Traffic accidents were not a problem in the pilot areas in United Kingdom either (in all areas there were 19 in the five years before and 1 after the redesign) [57], while in Vitoria-Gasteiz they were not monitored.

Despite the optimistic results from the studied areas, positive traffic calming effects on traffic safety reported from several other studies [23,60] and the improvement of traffic safety in Europe and Slovenia in general in recent years, the fact that settlements are still the most dangerous traffic areas is worrying. In 2017, 63% of all accidents and 35% of all fatal accidents in Slovenia occurred in settlements [68]. At European Union level, too, almost half of fatal traffic accidents occur in cities [69]. This indicates the need to change the approach to ensuring road safety. In recent years, a new paradigm has been introduced [18,33]. With it, the focus on motorized traffic, traffic safety and the reduction in driving speed typical of the classical paradigm has shifted to reducing the number of vehicles. New research shows that any use of motor vehicle poses a risk and that there are more fatal traffic accidents in environments with a higher number of kilometres driven [18,33]. Strategies to reduce car use and promote the use of alternative travel modes, such as sustainable urban mobility planning, thus reduce the risk of accidents and consequently increase road safety [62,70,71].

The effects of the redesign of the Juršovka neighbourhood in Ljutomer in line with the principles of comprehensive traffic calming and sustainable mobility paradigm [1] confirm the positive results of similar previous studies from the United Kingdom (2005) and Spain (2013) as well as from broader research on positive effects of traffic calming and measures improving conditions for sustainable mobility [1,23,40,56,58,72]. They prove that the effectiveness of such redesigns has not decreased over time. All the results show more active travel habits, higher quality of living, greater traffic safety, less motorized traffic and lower driving speeds, so it is unusual that the approach has not yet become a common design practice [60]. The planning process itself, which was highly transparent and inclusive and followed the key steps in the preparation of the SUMP, certainly contributed to the success of the measure [39]. The latter furthermore encouraged the Municipality of Ljutomer to define other areas of comprehensively calmed traffic in the municipality and to already order the production of detailed plans for four areas.

The development of a common methodology for monitoring the effects of comprehensive traffic calming would certainly be a welcome help and encouragement to other local communities, which would expand the knowledge base in this field by monitoring the effects of similar redesigns. An important factor for the wider use of comprehensive traffic calming is its placement and promotion within sustainable urban mobility planning and as a basic planning unit of SUMPs as was already promoted by the author and her colleagues from the Urban Planning Institute of the Republic of Slovenia [46]. Taking into account the already expressed need for coordinated packages of mutually reinforcing transport and land-use policies [2] and successful examples from Western Europe [56], it would also be necessary to explore the potential for its placement in traffic and spatial planning practice in Slovenia and other countries where traditional planning approaches still prevail.

## 5. Conclusions

Negative impacts of motor vehicle traffic in cities are still increasing despite the objectives that sustainable mobility paradigm put forward almost three decades ago. Measures to reduce them still primarily focus on traffic safety improvements through vehicle flows and speeds reduction (traffic calming). Comprehensive traffic calming targets the issue more comprehensively by also addressing changes in travel behaviour and quality of life and thus pursues objectives of sustainable mobility paradigm. Literature covering the effects of measures addressing all listed aspects is scarce.

Results of a longitudinal study presented in this paper show how the comprehensive traffic calming redesign of a residential neighbourhood impacts walking and cycling habits, quality of life, motorized traffic and road safety in a positive way. The studied redesign took place in 2016 while ex ante data gathering took place in 2014 and ex post data gathering took place in 2017 and 2018. Surveys and pilot interviews were used to consult residents; automatic pneumatic traffic counters were used to collect data on motorized traffic, and public databases were used to gather data on traffic accidents.

Around a third of residents claim to walk (37%), cycle (27%) and socialize (28%) more than before while around two thirds (63%) state that the quality of life in the neighbourhood has improved, and 75% strongly support the redesign elements. Moreover, vehicle speeds and flows and peak-hour flows have notably decreased throughout the neighbourhood. On average, the share of vehicles driving at less than 30 km/h increased by 20% (from 38% to 58%), mostly due to a reduction in the number driving between 30 and 50 km/h (59% before, 42% after). The share of vehicles exceeding 50 km/h has dropped to almost zero (3% before, 0.2% after). The 85th percentile speeds at a representative location were reduced by 6.97 km/h to 41.3 km/h. The average traffic flows on transit streets have decreased by 33% and on the access street by 2%. The average maximum peak hour flows have decreased by 30% (from 52 to 36 vehicles in the peak hour). Compared to the whole settlement of Ljutomer, very few traffic accidents occurred in the neighbourhood and only one after the redesign. All accidents were without casualties or serious injuries, and no pedestrians or cyclists were involved.

The results show a broad range of positive impacts of the comprehensive traffic calming approach and are confirmed by similar studies from the United Kingdom, Spain and elsewhere. Its potential for wider use in planning practice is substantial and should be promoted within the context of sustainable urban mobility planning and as a basic planning unit in SUMPs as is already the case in Slovenia. Furthermore, the development of a uniform approach to monitoring the effects of comprehensive traffic calming measures could also help to support and encourage other local authorities to monitor similar measures.

**Author Contributions:** Conceptualization, M.B. and M.L.; methodology, M.B.; investigation, M.B.; data curation, M.B.; writing—original draft preparation, M.B.; writing—review and editing, M.L.; visualization, M.B.; supervision, M.L. All authors have read and agreed to the published version of the manuscript.

**Funding:** This research received no external funding.

**Conflicts of Interest:** The authors declare no conflict of interest.

## References

1. Banister, D. The sustainable mobility paradigm. *Transp. Policy* **2008**, *15*, 73–80. [CrossRef]
2. Marshall, S. The challenge of sustainable transport.. In *Planning for a Sustainable Future*; Layard, A., Davoudi, S., Batty, S., Eds.; Spon Press: London, UK, 2001; pp. 131–147. ISBN 978-0-415-23227-2.

3. European Commission (EC). Together towards Competitive and Resource-Efficient Urban Mobility, COM 913 Final. Available online: https://ec.europa.eu/transport/themes/urban/urban_mobility/ump_en (accessed on 22 June 2018).
4. European Commission (EC). Action Plan on Urban Mobility, COM 490 Final. Available online: http://ec.europa.eu/transport/themes/urban/urban_mobility/action_plan_en.htm (accessed on 22 June 2018).
5. Rupprecht, S.; Brand, L.; Böhler-Baedeker, S.; Brunner, L.M. *Guidelines for Developing and Implementing a Sustainable Urban Mobility Plan*, 2nd ed.; Rupprecht Consult: Köln, Germany, 2019.
6. Plevnik, A.; Mladenovič, L.; Balant, M.; Koblar, S. *Nacionalni Program Celostnega Prometnega Načrtovanja 2018–2028*; Ministrstvo RS za infrastrukturo: Ljubljana, Slovenia, 2018.
7. Holden, E.; Gilpin, G.; Banister, D. Sustainable Mobility at Thirty. *Sustainability* **2019**, *11*, 1965. [CrossRef]
8. Freudendal-Pedersen, M.; Kesselring, S.; Servou, E. What is Smart for the Future City? Mobilities and Automation. *Sustainability* **2019**, *11*, 221. [CrossRef]
9. Næss, P.; Vogel, N. Sustainable urban development and the multi-level transition perspective. *Environ. Innov. Soc. Transit.* **2012**, *4*, 36–50. [CrossRef]
10. Freudendal-Pedersen, M. Mobility in daily life: Between freedom and unfreedom. In *Transport and Society*; Ashgate: Surrey, UK, 2009; ISBN 978-0-7546-7490-0.
11. European Commission (EC). White Paper. Roadmap to a Single European Transport Area–Towards a Competitive and Resource Efficient Transport System, COM 144 Final. Available online: http://ec.europa.eu/transport/themes/strategies/2011_white_paper_en.htm (accessed on 22 June 2018).
12. European Commission (EC). A European Strategy for Low-Emission Mobility, COM 501 Final. Available online: https://ec.europa.eu/transport/sites/transport/files/themes/strategies/news/doc/2016-07-20-decarbonisation/com%282016%29501_en.pdf (accessed on 22 June 2018).
13. European Transport Safety Council (ETSC). Available online: www.etsc.eu (accessed on 1 May 2020).
14. Plevnik, A. *Okolje, Promet in Zdravje*; Agencija RS za okolje: Ljubljana, Slovenia, 2016.
15. Welle, B.; Wei, L.; Adriazola, C.; King, R.; Obelheiro, M.; Sarmiento, C.; Liu, Q. *Cities Safer by Design: Guidance and Examples to Promote Traffic Safety Through Urban and Street Design: Version 1.0*; World Resource Institute: Washington, DC, USA, 2015; ISBN 978-1-56973-866-5.
16. World Health Organisation (WHO). *Physical Activity, Key Facts*. Available online: www.who.int/news-room/fact-sheets/detail/physical-activity (accessed on 4 April 2020).
17. World Health Organisation (WHO). Available online: www.who.int (accessed on 1 May 2020).
18. Litman, T. *A New Traffic Safety Paradigm*; Victoria Transport Policy Institute: Victoria, BV, Canada, 2020; p. 45.
19. Plevnik, A.; Balant, M.; Mladenovič, L. Skrb vzbujajoče spremembe v mobilnosti mladih–primer osnovnošolcev v Novem mestu. *Urbani Izziv* **2017**, *28*, 70–79. [CrossRef]
20. De Nazelle, A.; Nieuwenhuijsen, M.J.; Antó, J.M.; Brauer, M.; Briggs, D.; Braun-Fahrlander, C.; Cavill, N.; Cooper, A.R.; Desqueyroux, H.; Fruin, S.; et al. Improving health through policies that promote active travel: A review of evidence to support integrated health impact assessment. *Environ. Int.* **2011**, *37*, 766–777. [CrossRef]
21. Mueller, N.; Rojas-Rueda, D.; Cole-Hunter, T.; de Nazelle, A.; Dons, E.; Gerike, R.; Götschi, T.; Int Panis, L.; Kahlmeier, S.; Nieuwenhuijsen, M. Health impact assessment of active transportation: A systematic review. *Prev. Med.* **2015**, *76*, 103–114. [CrossRef]
22. Kelly, P.; Kahlmeier, S.; Götschi, T.; Orsini, N.; Richards, J.; Roberts, N.; Scarborough, P.; Foster, C. Systematic review and meta-analysis of reduction in all-cause mortality from walking and cycling and shape of dose response relationship. *Int. J. Behav. Nutr. Phys. Act.* **2014**, *11*, 132. [CrossRef]
23. Pucher, J.; Buehler, R. Making Cycling Irresistible: Lessons from The Netherlands, Denmark and Germany. *Transp. Rev.* **2008**, *28*, 495–528. [CrossRef]
24. Leeuw, E.; De Tsouros, A.D.; Dyakova, M.; Green, G. *Healthy Cities, Promoting Health and Equity-Evidence for Local Policy and Practice: Summary Evaluation of Phase V of the WHO European Healthy Cities Network*; World Health Organisation Regional Office for Europe: Copenhagen, Denmark, 2014; ISBN 978-92-890-5069-2.
25. State of Green. *Sustainable Urban Transportation. Creating Green Liveable Cities*; Version 1.0.; State of Green: Çopenhagen, Denmark, 2016.
26. European Commission (EC). Urban Agenda for the EU. Partnership for Urban Mobility. Developing Guidelines on Infrastructure for Active Mobility Supported by Relevant Funding. Available online: https://ec.europa.eu/futurium/en/system/files/ged/final_deliverable_action_5.pdf (accessed on 22 September 2020).

27. Carmona, M.; Heath, T.; Oc, T.; Tiesdell, S. *Public Places-Urban Spaces: The Dimensions of Urban Design*, 2nd ed.; Carmona, M., Ed.; Architectural Press: Oxford, UK; Elsevier: Amsterdam, The Netherlands, 2010; ISBN 978-1-85617-827-3.
28. Carmona, M.; Gabrieli, T.; Hickman, R.; Laopoulou, T.; Livingstone, N. Street appeal: The value of street improvements. *Prog. Plan.* **2018**, *126*, 1–51. [CrossRef]
29. Gehl, J. *Cities for People*; Island Press: Washington, DC, USA, 2010; ISBN 978-1-59726-573-7.
30. Gehl, J. *Life between Buildings: Using Public Space*; Island Press: Washington, DC, USA, 2011; ISBN 978-1-59726-827-1.
31. Jones, P.; Marshall, S.; Boujenko, N. Creating More People-Friendly Urban Streets Through 'Link and Place' Street Planning and Design. *IATSS Res.* **2008**, *32*, 14–25. [CrossRef]
32. Lauwers, D. Functional road categorization: New concepts and challenges related to traffic safety, traffic management and urban design-reflections based on practices in Belgium confronted with some Eastern European cases. In Proceedings of the Transportation and Land Use Interaction, Bucharest, Hungary, 23–25 October 2008; Volume 8, pp. 149–164.
33. Litman, T. *Evaluating Transportation Diversity: Multimodal Planning for Efficient and Equitable Communities*; Victoria Transport Policy Institute: Victoria, BC, Canada, 2017; p. 41.
34. University of Leeds KonSULT. The Knowledgebase on Sustainable Urban Land Use and Transport. Available online: http://www.konsult.leeds.ac.uk/ (accessed on 31 July 2017).
35. Vivanco, A.A.; Escudero, J.C. *The Sustainable Urban Mobility Plan of Vitoria Gasteiz, Spain*, CIVITAS PROSPERITY Project. 2017.
36. Escudero, J.C. *Innovation brief on Superblocks*, CIVITAS PROSPERITY Project. 2017.
37. Goeminne, J.; Hanssens, C. *The Sustainable Urban Mobility Plan of Sint Niklaas, Belgium*, CIVITAS PROSPERITY Project. 2017.
38. Ajuntament de Barcelona. Let's fill streets with life: Establishing Superblocks in Barcelona. Ajuntament de Barcelona, Commission for Ecology, Urban Planning and Mobility. 2016.
39. ISIS. *Civitas Modern. Final Evaluation Report*, CIVITAS MODERN Project. 2013.
40. De Geest, L. *Gent's Traffic Circulation Plan*. ELTIS. Available online: www.eltis.org/discover/case-studies/gents-traffic-circulation-plan-belgium (accessed on 11 August 2020).
41. Ministrstvo RS za infrastrukturo (MZI). *Infrastruktura za pešce. Splošne usmeritve. Verzija 1.0*; Ministrstvo RS za Infrastrukturo: Ljubljana, Slovenia, 2017.
42. Bartle, C.; Calvert, T.; Clark, B.; Hüging, H.; Jain, J.; Melia, S.; Mingardo, G.; Rudolph, F.; Ricci, M.; Parkin, J.; et al. *The Economic Benefits of Sustainable Urban Mobility Measures. Independent Review of Evidence: Reviews*; EVIDENCE Project: Brussels, Belgium, 2016.
43. Shergold, I.; Parkhurst, G. *The Economic Benefits of Sustainable Urban Mobility Measures. Independent Review of Evidence: Method*; EVIDENCE Project: Brussels, Belgium, 2016.
44. Brown, V.; Moodie, M.; Carter, R. Evidence for associations between traffic calming and safety and active transport or obesity: A scoping review. *J. Transp. Health* **2017**, *7*, 23–37. [CrossRef]
45. Plevnik, A.; Mladenovič, L.; Balant, M.; Ružić, L. *Prijazna mobilnost za zadovoljno prihodnost: Prometna strategija Občine Ljutomer*; Ministrstvo RS za infrastrukturo in prostor: Ljubljana, Slovenia, 2012.
46. Balant, M.; Plevnik, A. People friendly travel zones as a basic planning unit in SUMPs—Case study from Ljutomer. In Proceedings of the 6th International Conference "Towards a Humane City", Novi Sad, Serbia, 26–27 October 2017; pp. 149–164.
47. Sheller, M.; Urry, J. The New Mobilities Paradigm. *Environ. Plan. A* **2006**, *38*, 207–226. [CrossRef]
48. Ščetinin, V.; Balant, M.; Plevnik, A.; Mladenovič, L. *Ureditev Stanovanjske Soseske Juršovka kot Območja Prijaznega Prometa, Projektna Dokumentacija IDP*; Biro skiro: Ljubljana, Slovenia, 2014.
49. CIVITAS PROSPERITY Project. Available online: http://sump-network.eu/ (accessed on 7 August 2020).
50. Lep, M.; Turnšek, S.; Vodnik, S. *Prometna študija: Izvedba štetja prometa, meritev hitrosti in analiza rezultatov v Občini Ljutomer*; Fakulteta za gradbeništvo, Univerza v Mariboru: Maribor, Slovenia, 2014.
51. Lep, M.; Turnšek, S. *Prometna študija: Izvedba štetja prometa, meritev hitrosti in analiza rezultatov v Občini Ljutomer*; Fakulteta za gradbeništvo, prometon inženirstvo in arhitekturo, Univerza v Mariboru: Maribor, Slovenia, 2017.

52. Lep, M.; Turnšek, S. *Prometna študija: Izvedba štetja prometa, meritev hitrosti in analiza rezultatov v Občini Ljutomer*; Fakulteta za gradbeništvo, prometno inženirstvo in arhitekturo, Univerza v Mariboru: Maribor, Slovenia, 2018.
53. Ministrstvo RS za Notranje Zadeve (MNZ): Prometna Varnost, Statistične Datoteke. Available online: https://www.policija.si/o-slovenski-policiji/statistika/prometna-varnost (accessed on 24 July 2020).
54. Urry, J. The 'System' of Automobility. *Theory Cult. Soc.* **2004**, *21*, 25–39. [CrossRef]
55. Appleyard, D. *Livable Streets*; University of California Press: Berkeley, CA, USA, 1981; ISBN 978-0-520-03689-5.
56. Buehler, R.; Pucher, J.; Gerike, R.; Götschi, T. Reducing car dependence in the heart of Europe: Lessons from Germany, Austria, and Switzerland. *Transp. Rev.* **2017**, *37*, 4–28. [CrossRef]
57. Webster, D.; Tilly, A.; Wheeler, A.; Nicholls, D.; Buttress, S. *Pilot home zone schemes: Summary of the schemes, prepared for Traffic Management Division, Department for Transport*; Webster, D., Ed.; TRL report; TRL Limited: Crowthorne, UK, 2005; ISBN 978-1-84608-653-3.
58. CIVITAS POINTER. *Superblocks Model. Vitoria-Gasteiz*, CIVITAS POINTER Project, CIVITAS POINTER Project. 2013.
59. Vaitkus, A.; Čygas, D.; Jasiūnienė, V.; Jateikienė, L.; Andriejauskas, T.; Skrodenis, D.; Ratkevičiūtė, K. Traffic Calming Measures: An Evaluation of the Effect on Driving Speed. *PROMET* **2017**, *29*, 275–285. [CrossRef]
60. Ewing, R.; Dumbaugh, E. The Built Environment and Traffic Safety: A Review of Empirical Evidence. *J. Plan. Lit.* **2009**, *23*, 347–367. [CrossRef]
61. Elvik, R. Transportøkonomisk institutt (Norway). In *The Power Model of the Relationship between Speed and Road Safety: Update and New Analyses*; Transportøkonomisk Institutt: Oslo, Norway, 2009; ISBN 978-82-480-1002-9.
62. Elvik, R.; Bjørnskau, T. Safety-in-numbers: A systematic review and meta-analysis of evidence. *Saf. Sci.* **2017**, *92*, 274–282. [CrossRef]
63. Richter, E.D.; Berman, T.; Friedman, L.; Ben-David, G. Speed, road injury, and public health. *Annu. Rev. Public Health* **2006**, *27*, 125–152. [CrossRef]
64. Bunn, F.; Collier, T.; Frost, C.; Ker, K.; Steinbach, R.; Roberts, I.; Wentz, R. Area-wide traffic calming for preventing traffic related injuries. *Cochrane Database Syst. Rev.* **2003**. [CrossRef]
65. Elvik, R. Area-wide urban traffic calming schemes: A meta-analysis of safety effects. *Accid. Anal. Prev.* **2001**, *33*, 327–336. [CrossRef]
66. Jones, P.; Boujenko, N. Link and Place: A New Approach to Street Planning and Design. *Road Transp. Res.* **2009**, *18*, 38–48.
67. Agerholm, N.; Knudsen, D.; Variyeswaran, K. Speed-calming measures and their effect on driving speed–Test of a new technique measuring speeds based on GNSS data. *Transp. Res. Part. F: Traffic Psychol. Behav.* **2017**, *46*, 263–270. [CrossRef]
68. Javna agencija RS za varnost prometa (AVP). *Analiza in Pregled Stanja Varnosti v Cestnem Prometu za leto 2017*; Javna Agencija RS za varnost prometa: Ljubljana, Slovenia, 2018; p. 13.
69. European Commission (EC). On the Implementation of Objective 6 of the European Commission's Policy Orientations on Road Safety 2011–2020—First Milestone Towards an Injury Strategy, SWD 94 final. 2013. Available online: https://ec.europa.eu/transport/road_safety/sites/roadsafety/files/pdf/ser_inj/ser_inj_swd.pdf (accessed on 22 June 2018).
70. Jacobsen, P.L. Safety in numbers: More walkers and bicyclists, safer walking and bicycling. *Inj. Prev.* **2003**, *9*, 205–209. [CrossRef] [PubMed]
71. Fyhri, A.; Sundfør, H.B.; Bjørnskau, T.; Laureshyn, A. Safety in numbers for cyclists-conclusions from a multidisciplinary study of seasonal change in interplay and conflicts. *Accid. Anal. Prev.* **2016**, *105*, 124–133. [CrossRef] [PubMed]
72. Jones, P.; Boujenko, N.; Marshall, S. *Link & Place: A Guide to Street Planning and Design*; Local Transport Today Ltd.: London, UK, 2007; ISBN 978-1-899650-41-5.

Article

# Assessing Sustainable Mobility Measures Applying Multicriteria Decision Making Methods

**Jonas Damidavičius [1,*], Marija Burinskienė [1] and Jurgita Antuchevičienė [2]**

1 Department of Roads, Vilnius Gediminas Technical University, Sauletekio al. 11, 2510 Vilnius, Lithuania; marija.burinskiene@vgtu.lt

2 Department of Construction Management and Real Estate, Vilnius Gediminas Technical University, Sauletekio al. 11, 2510 Vilnius, Lithuania; jurgita.antucheviciene@vgtu.lt

* Correspondence: jonas.damidavicius@vgtu.lt; Tel.: +370-685-53836

Received: 3 June 2020; Accepted: 27 July 2020; Published: 28 July 2020

**Abstract:** An increasing number of recent discussions have focused on the need for designing transport systems in consonance with the importance of the environment, thus promoting investment in the growth of non-motorized transport infrastructure. Under such conditions, the demand for implementing the most effective infrastructure measures has a profoundly positive impact, and requires the least possible financial and human resources. The development of the concept of sustainable mobility puts emphasis on the integrated planning of transport systems, and pays major attention to the expansion of non-motorized and public transport, and different sharing systems, as well as to effective traffic management involving intelligent transport systems. The development of transport infrastructure requires massive investment, and hence the proper use of mobility measures is one of the most important objectives for the rational planning of sustainable transport systems. To achieve this established goal, this article examines a compiled set of mobility measures and identifies the significance of the preferred tools, which involve sustainable mobility experts. The paper also applies multicriteria decision making methods in assessing urban transport systems and their potential in terms of sustainable mobility. Multicriteria decision making methods have been successfully used for assessing the effectiveness of sustainable transport systems, and for comparing them between cities. The proposed universal evaluation model is applied to similar types of cities. The article explores the adaptability of the model by assessing big Lithuanian cities.

**Keywords:** sustainable urban mobility; SUMP; mobility measures; multicriteria decision making methods; MCDM

## 1. Introduction

The adverse effect of transport on the environment is currently being addressed through the development of transport infrastructure in cities that have not preserved comprehensive traditions in sustainable mobility planning, thus further encouraging the use of private cars, increasing congestion on roads and causing plenty of negative consequences like loss of time, transport pollution and traffic accidents. The outlined situation has arisen mainly due to urban sprawl, because transport systems are not adapted to the needs of all age, social and interest groups, and no alternatives to travelling by car have been proposed.

In recent years, under the guidance of the White Paper 'Roadmap to a Single European Transport Area—Towards a competitive and resource efficient transport system' (White Paper on Transport) [1] and the Communication from the Commission to the European Parliament, the Council, the European Economic and Social Committee and the Committee of the Regions 'Together Towards Competitive and Resource-Efficient Urban Mobility' (Communication) [2], a solid theoretical basis for preparing sustainable urban mobility plans (SUMP) has emerged in Europe. Long-term integrated thinking in planning Sustainable Urban Mobility (SUM) systems is one of the most important tasks that must occur in the daily activities of all stakeholders.

The key principles for successful SUM cover the involvement of the public and stakeholders in planning and implementation processes, promoting institutional cooperation on transport links in order to deal with the issues of the other aspects of urban life, identifying the most effective urban infrastructure and sustainable mobility measures (hereinafter referred to as mobility measures), and monitoring and assessing mobility measures and the implementation process.

A huge number of EU-funded projects and programs have provided valuable knowledge that has helped cities to take a new approach to urban planning and transport infrastructure. Modern SUM planning is increasingly evaluated as a necessity for European cities looking towards a common future.

SUM planning is a strategic and integrated method for effectively dealing with the issues of urban transportation. The main goal of the method is to improve accessibility and life quality by switching to alternative transport. This technique is based on decision-making in the long run, which requires a thorough assessment of the current situation and future trends, a mutually acceptable vision and strategic goals, as well as a set of integrated measures (regulatory, promotional, financial, technical and infrastructural) for achieving the established goals. The imposed measures should be regularly monitored and assessed [3].

Hence, with reference to the detailed analysis of practical research and methods applied in other countries, this study aims to reasonably classify mobility measures and assess their significance in line with the size and characteristics of the city. Also, multicriteria decision making methods (MCDM) assist in assessing the transport systems of the biggest Lithuanian cities in terms of sustainable mobility.

Section 2 presents the study process, Section 3 describes the methodology for setting up the package of mobility measures, Section 4 identifies the significance of mobility measures, Section 5 presents the results of MCDM evaluation and Section 6 sets out conclusions and insights.

## 2. The Research Process

Pursuant to the White Paper on Transport and Communication, in 2015, national guidelines for developing Sustainable Urban Mobility Plans were approved in Lithuania [4]. The guidelines covered nine thematic areas that were recommended for further development. The included promotion of public transport (T1), non-motor vehicle integration (T2), traffic safety and transport security (T3), improvement to traffic organization and mobility management (T4), city logistics (T5), integration of people with special needs (T6), promotion of alternative fuels and clean vehicles (T7), assessment of demand for intelligent transport systems (T8) and modal shift (T9). The mobility measures further described in the article are also classified, conforming to the above-introduced thematic areas, but excluding the modal shift, which is more frequently expressed to estimate the impact or result achieved than to describe a set of the specified mobility measures.

The application of MCDM runs into problems because each method gives different meanings in the context of the same criteria. Thus, we usually see to the integrated application of MCDM methods, whereby the results of different MCDM methods are analyzed employing such techniques as the Weighted average, Borda or Copeland, which summarize the findings.

The article is aimed at assessing the largest cities of Lithuania in terms of sustainable mobility. To achieve this objective, the following tasks have been set:

- compiling the commonly used sets of mobility measures considering thematic areas;
- compiling expertise to determine the relevance of mobility measures;
- applying MCDM methods in the assessment of cities;
- analyzing findings using the Weighted average (WAM), Borda and Copeland methods.

The research process is provided in Figure 1.

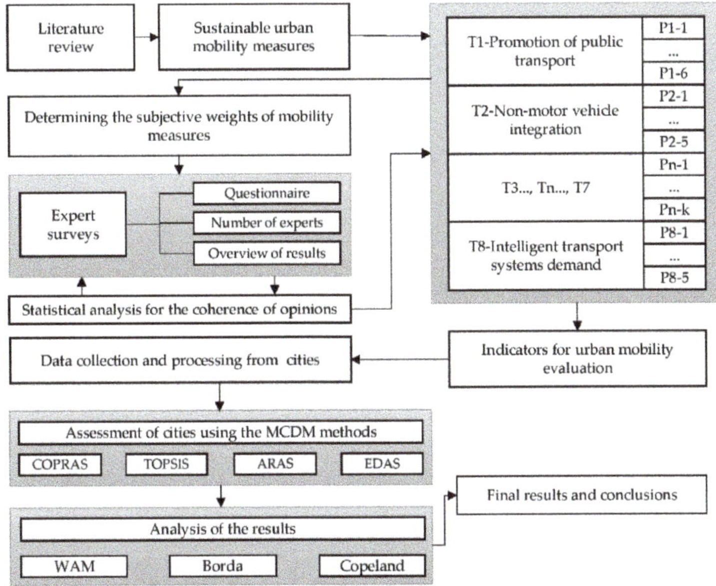


**Figure 1.** City assessment applying MCDM methods.

## 3. Designing a System of Mobility Measures

Mobility measures cover a wide range of instruments for achieving the objectives of sustainable development inside the transport sector, and tools for solving the identified transport problems [5]. The identification of efficient mobility measures is the basis of planning SUM.

Until now, the scientific literature has not provided extensive research on SUM development and its impact on urban population, which has been a consequence of the transport infrastructure being designed for the ease of use of road vehicles for a long time. Thus, only street network densification, street widening and traffic throughput were considered [3]. However, in line with the White Paper on Transport, the approach to planning transport systems has changed, and is now more focused on the mobility of people and the effective operation of the infrastructure and transport services that they require.

To solve this issue, researchers, transport experts, the representatives of local authorities and various research agencies have started designing several systems for assessing SUM efficiency and cost-effectiveness, in order to identify the mobility measures that have the greatest positive impact and their economic benefits. The use of sustainable mobility measures takes many forms, some of which see the country-specific manifestation of the mobility measures having the greatest impact [6–18], while others develop a mobility index based on mobility measures [19–26], or assess sustainable mobility through the environmental, economic and social prism [14,20,27–29].

The process of selecting mobility measures is clearly described in the Guidelines for Developing and Implementing a Sustainable Urban Mobility Plan (Second Edition) (Figure 2) [3].

The development of transport infrastructure and the implementation of mobility measures requires a large budget, which often becomes a serious problem for urban governance and, as a result, the development process becomes very slow. The effective and rational selection of mobility measures is a prerequisite for adhering to the principles of economy and acceptability. The assessment and selection of mobility measures needs to identify the most appropriate and cost-effective tools for the chosen development scenario. In order to assess all available options, a comprehensive long list of mobility measures should first be established, which should be based on individual and local expertise, stakeholder and societal ideas, the experience of professionals from other cities and the databases of mobility measures.

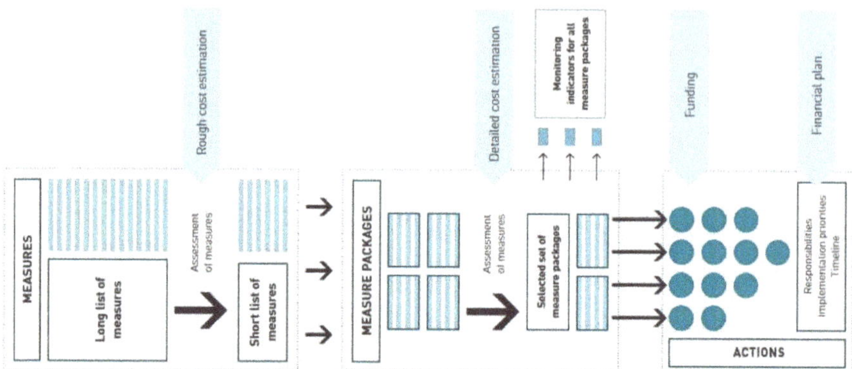


**Figure 2.** The process of selecting mobility measures [3].

Implementing mobility measures in isolation will not achieve the set goals and objectives, and therefore classifying measures and forming their packages is required [3,12,30]. In designing SUM development plans, it is necessary to consider how different mobility measures interact with each other to create a better result than those implemented individually. The creation of such sets is often referred to as an integrated approach, or the integrated implementation of mobility measures. In the development of sets of mobility measures, we often find that relevant identified mobility measures working together have greater impacts (synergy of mobility measures), or they may be designed to enhance the effectiveness of other measures (complementary).

A set of mobility measures is a combination of complementary mobility measures that are often attributed to different categories, and are well-coordinated so as to address specific challenges and overcome barriers to their implementation more effectively than individual mobility measures [5]. In order to create the most useful sets of mobility measures, different ways of grouping should be explored and tested.

T. Litman reviewed and researched many different systems of sustainable transport indicators [31] (revised 54 transport sustainability assessment systems applied in 22 countries), and arrived at the very interesting and useful conclusion that sustainable mobility indicators could not always properly assess urban SUM. For example, if the application area of an indicator is very narrow, it does not reflect the true value with regard to SUM (if only the development of vehicles using an ecological fuel is assessed, refusing the assessment of traffic congestion and road accidents, the assessment indicator does not reflect the real situation). The same is true if intermediate targets, rather than the final result, are considered (the length of cycling routes is an intermediate result, which may not necessarily correspond with the final result, which is a larger number of users). The same principle should apply to the selection and planning of mobility measures: the overall SUM effect requires an integrated approach, rather than the implementation of individual mobility measures.

The search for effective mobility measures and their sets is an increasingly important topic among spatial planning, mobility and transport experts. Therefore, in recent years, many international projects aimed at listing effective mobility measures in relation to certain development criteria or directions have been implemented. Much attention has been paid to developing sets of mobility measures through the CH4LLENGE project [32], which assessed and presented 64 different types of mobility measures. Each of them can be implemented in different ways, subject to the local needs, and thus many possible combinations of the mobility measures present in the sets are available. The mobility measures proposed by the project were integrated into the KonSULT interactive database [11], aimed at helping policy makers, practitioners, experts and other stakeholders to understand urban mobility issues, and identify relevant mobility measures and packages. With reference to the results and recommendations of the CH4LLENGE project, the SUMPs-Up project [6–8] provides a long list of mobility measures (more than 100) that fall into 25 categories, related to three types of cities with different levels of SUM development: beginners, advanced and developed. Within the scope of the project, SUM development recommendations for these types of cities were issued. As for other projects, for example EVIDENCE [9], the mobility measures that would have the greatest economic impact were examined. As a result, a list of 22 most cost-effective mobility measures was drawn up, and a detailed description was prepared for each mobility measure, thus indicating the application, implementation and expected cost-effectiveness of the mobility measure.

Experts from the international management consulting company Arthur D Little, and the International Association of Public Transport (UITP), devised an assessment system for urban transport services, composed of 19 key mobility measures each rated with a certain score [15]. A similar assessment was devised by Costa [25], who created the Sustainable Urban Mobility Index (SUMI)—a tool for assessing SUM based on the multicriteria approach. The SUMI was made of 87 indicators proposed by [33]. The indicators were carefully selected to reflect a diverse impact and the outlooks of SUM. J. Lima et al. [21] used both the SUMI and the SUM development method proposed by Mancini [23], who analyzed mobility measures under three categories: cost, time required for implementation, and political risk in the implementation process.

Reisi et al. [20] explored a large quantity of scientific literature, examining the development of SUM-creating systems for indicators and mobility measures. The scientists created an individual SUM assessment index that consisted of nine mobility measures, divided in line with the principles of sustainable development, including environmental, social and economic aspects. To compile the index, mobility measures were assigned assessment indicators for their implementation. The indicators were attributed via their determined significance. In agreement with the principles of sustainable development, mobility measures were also assessed by T. Shiau et al. [16], who used the Rough sets theory [34] and identified the 26 mobility measures that have the greatest impact on SUM. T. Shiau and J. Liu [28] grouped the system of mobility measures conforming to economic, environmental, social and energy aspects, and determined the significance of the mobility measures using the analytical hierarchy process (AHP). Burinskienė et al. [27] assessed a list of 38 mobility measures pursuant to the principles of sustainable development, thus assigning a higher or lower significance to the impacts of each of the measures.

Following a revision of the research literature (the assessment of the prepared international projects containing the sets of the compiled mobility measures and the process of selecting mobility measures), presented in Figure 2, the authors selected 38 mobility measures and divided them into eight thematic areas.

For the further analysis of mobility measures, expertise is used, which allows us to summarize the opinions of the expert group so as to devise a possible solution to the problem.

## 4. Determining the Significance of Mobility Measures

The Guidelines for Developing and Implementing a Sustainable Urban Mobility Plan (Second Edition) [3] place more emphasis on planning norms, and make recommendations for highlighting city size and its characteristics, specificity, planning differences, urban management, and different integration practices of transport modes and travel habits. This is also important for determining the relevance of mobility measures, because different types of cities have different needs as regards mobility measures. In order to properly establish the significance of mobility measures, the authors divided Lithuanian cities with SUMPs according to population and functional purpose (Figure 3).

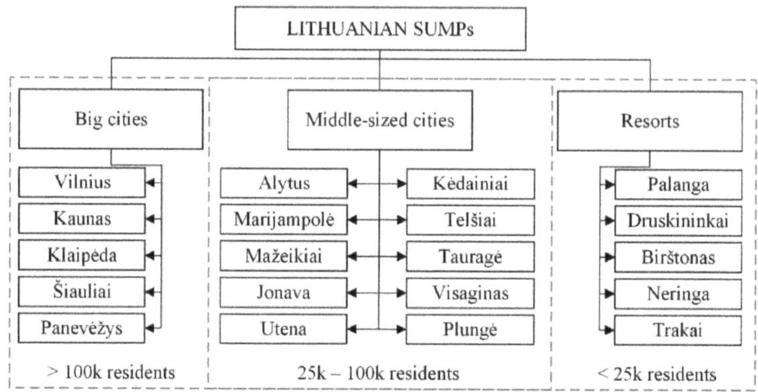


**Figure 3.** The categories of cities according to population and functional purpose.

Big cities (more than 100 thousand inhabitants) are the Lithuanian cities that are most characterized by a growing economy and a large supply of jobs and services. These cities are strong regional centers, daily attracting a large workforce. The population remains stable, or is slightly increasing. Heavy vehicle traffic predominates in these areas, congestion is frequent during rush hour, and many trips, a wide range of transport options, and a high need for infrastructure development and renovation or repair are observed. These cities are more likely to apply solutions produced by intelligent transport systems, and generate high economic demands.

Middle-sized cities (from 25 thousand to 100 thousand inhabitants) are frequently referred to as metropolitan satellite towns (residential areas). They are often the centers of smaller regions or industrial cities that provide jobs for the local people. The populations remain stable or decline slightly, due to constant migration to nearby cities. These areas are predominated by moderate traffic, infrequent congestion during rush hour, and well-developed infrastructure, which requires less investment in development but more in renovation or repair. A strong focus is placed on the development and maintenance of public spaces. Low supply and demand for transport innovation is observed.

Resorts (up to 25 thousand inhabitants) are strongly characterized by seasonality, when both vehicle flow and the number of city newcomers increase. These towns are characterized by recreational and entertainment services. Visitors prefer these places because of their natural diversity and the quality of the services provided. The local population is not very large, ranging from 3 to 25 thousand. Transport is well developed, with a primary focus on non-motorized transport infrastructure. There is a large supply of transport sharing and rental.

The article discusses and assesses the significance of mobility measures in Lithuania's biggest cities using MCDM methods. The same evaluation process was carried out for middle-sized cities and resorts, but due to the large volume of data, the article provides only the results from the evaluation of the biggest cities in Lithuania.

The relevance of mobility measures was assessed by interviewing sustainable mobility and transport experts, academicians, advisors, and the representatives of municipal administrations, stakeholder institutions and Non-governmental organizations that have experience in dealing with sustainable mobility and urban planning, and implementing infrastructure measures. The expert survey was based on the separate evaluation of thematic areas, and on the case-by-case assessment of mobility measures for each thematic area, in relation to the different types of cities.

For determining the significance of the impact of mobility measures in big cities, 19 experts were surveyed. Of those experts, 63% were involved in the development of sustainable mobility plans, 47% of the respondents implemented sustainable mobility plans, 32% of the experts participated in sustainable mobility education and policy-making, and 16% of the respondents took part in academic activities. The survey showed that the selected respondents were most frequently involved in more than one area of sustainable mobility, i.e., an expert was engaged in the development of sustainable mobility plans and educational activities, and therefore their interest and experience could be expected to provide high-quality and representative assessment.

The surveyed experts were asked to assess thematic areas and mobility measures imposed in line with the presented thematic areas (hereinafter—assessment indicators), and answer the following questions:

- What thematic areas are the most important for developing SUM?
- What mobility measures have the largest impact on SUM development?

Assessment indicators were ranked pursuant to importance. The indicator with the highest value was assessed as having the biggest possible score, and the indicator with the lowest value was given the lowest score. The peer review showed expert preference for thematic areas and the mobility measures of individual thematic areas. With reference to the ranks given, the weights of the mobility measures were determined. The expert survey assessed the consistency of expert opinions [35–39].

First, the sum of ranks for each assessment indicator is determined:

$$P_j = \sum\nolimits_{k=1}^{r} P_{jk} \tag{1}$$

where $P_{jk}$ is the expert rank $k$ for assessment indicator $j$, and $r$ is the number of experts.

The subjective weight of the index is equal to:

$$\omega_j = \frac{\overline{P}_j}{\sum_{i=1}^{m} P_j} \tag{2}$$

where $m$ is the number of assessment indicators.

The rating of the index enables us to verify the agreement among expert opinions. Kendall's coefficient of concordance, $W$, determines the agreement level, calculated via the following formula [35]:

$$W = \frac{12 \cdot S}{r^2 \cdot m \cdot (m^2 - 1)} \tag{3}$$

The sum of squared deviations, S, of ranking sums' $P_j$ deviations from the total mean P is calculated via the following formula:

$$S = \sum\nolimits_{j=1}^{m} (P_j - \overline{P})^2 \tag{4}$$

The level of expert agreement is determined by the related value of $\chi^2$, rather than by the coefficient of concordance $W$, and is calculated as reported in formula [35]:

$$\chi^2 = W \cdot r \cdot (m-1) = \frac{12S}{r \cdot m \cdot (m+1)} \tag{5}$$

The statistical hypothesis concerning the expert agreement on ranks has been proven [35] to be acceptable for making calculations as presented in the last formula, where the value of $\chi^2$ is higher than the critical value of ${\chi^2}_{kr}$ taken from the $\chi^2$ distribution table, with the freedom degree equal to $v = m - 1$ and the selected significance level $\alpha$ close to zero.

The calculated significances of the mobility measures indicate the instances where the selected assessment indicator is more important than the other assessment indicator. Thus, the following selections of assessment indicators involve only the indicators with the highest values in each of the intended groups.

The parameters and determined significance from the expert survey of the assessment indicators are provided in Table 1.

**Table 1.** The parameters and significance of the expert survey of assessment indicators.

| Code | Mobility Measure | Weight (ω) | Rank | Code | Mobility Measure | Weight (ω) | Rank |
|---|---|---|---|---|---|---|---|
| T1 | Promotion of public transport | 0.1813 | 1 | P4-1 | Awareness Campaign, Events and Promotional Activities | 0.2561 | 1 |
| T2 | Non-motor vehicle integration | 0.1389 | 4 | P4-2 | Car parking management | 0.1544 | 4 |
| T3 | Traffic safety and transport security | 0.1623 | 2 | P4-3 | Parking Charges | 0.2421 | 2 |
| T4 | Improvement to traffic organization and mobility management | 0.1477 | 3 | P4-4 | Car Sharing | 0.1368 | 5 |
| T5 | City logistics | 0.0863 | 7 | P4-5 | Park and Ride | 0.2105 | 3 |
| T6 | Integration of people with special needs | 0.1155 | 5 | P5-1 | A driving ban for lorries | 0.1789 | 4 |
| T7 | Promotion of alternative fuels and clean vehicles | 0.0716 | 8 | P5-2 | Urban Consolidation Centers | 0.1895 | 3 |
| T8 | Assessment of demand for Intelligent transport systems | 0.0965 | 6 | P5-3 | Access restrictions | 0.3368 | 1 |
| P1-1 | Conventional Timetable | 0.2080 | 1 | P5-4 | New road construction | 0.2947 | 2 |
| P1-2 | Public transport priority lanes | 0.1880 | 3 | P6-1 | Mobility infrastructure for people with disabilities | 0.3211 | 1 |
| P1-3 | Public transport tickets and fare levels | 0.0952 | 6 | P6-2 | Accessibility of the main transport points | 0.2316 | 3 |
| P1-4 | Rapid public transport transit | 0.1830 | 4 | P6-3 | Shared spaces | 0.1737 | 4 |
| P1-5 | Public transport Terminals and Interchanges/Stops | 0.1203 | 5 | P6-4 | Accessible public transport | 0.2737 | 2 |
| P1-6 | Public transport network | 0.2055 | 2 | P7-1 | Alternative fuel public transport | 0.3053 | 2 |
| P2-1 | Pedestrian routes, networks | 0.2596 | 2 | P7-2 | Alternative fuel supply infrastructure | 0.1842 | 4 |
| P2-2 | Cycle Networks | 0.2632 | 1 | P7-3 | Low-emission zones | 0.3158 | 1 |
| P2-3 | Cycle Parking and Storage | 0.1474 | 4 | P7-4 | Promotion of alternative fuel vehicles | 0.1947 | 3 |
| P2-4 | Bike Sharing | 0.1333 | 5 | P8-1 | Intelligent traffic light system | 0.2351 | 2 |
| P2-5 | Lighting the cycle and pedestrian network | 0.1965 | 3 | P8-2 | Integrated Ticketing | 0.1965 | 3 |
| P3-1 | Traffic cameras | 0.1509 | 4 | P8-3 | Bus priority intersections | 0.2526 | 1 |
| P3-2 | Safety intersections | 0.2456 | 2 | P8-4 | Real Time Passenger Information | 0.1544 | 5 |
| P3-3 | Safety pedestrian and cycling crossing facilities | 0.2596 | 1 | P8-5 | Congestion charging | 0.1614 | 4 |
| P3-4 | Road Maintenance | 0.1228 | 5 | | | | |
| P3-5 | Low speed zones | 0.2211 | 3 | | | | |

The established significance values of the assessment indicators are further used for the multicriteria assessment of thematic areas using MCDMs.

## 5. The Results of Applying Multicriteria Decision Making Methods

Many MCDMs can be applied in analyzing the implementation of urban mobility measures in cities. Parezanovic et al. [40] used the COPRAS method for assessing mobility measures in line with the selected assessment criteria. Hickman et al. [41] applied multicriteria assessment in exploring the possibilities of developing transport infrastructure in keeping with different scenarios. Podvezko V. and Sivilevičius H. [42] employed the AHP method and examined transport systems through the prism of traffic safety. Other researchers applied MCDM methods in specifying locations to implement mobility measures [43].

MCDM are aimed at creating cumulative indicators for each selected thematic area. The indicator reflects the attractiveness of the area in quantitative measures, expressed in the unit value [39,44,45], which means that the calculated value defines the attractiveness of the thematic area for urban development.

The past conducted research has applied the COPRAS (Complex Proportional Assessment) method in combining the values of all the mobility measures into a single qualitative assessment—the value of the indicator of the method [46]. The TOPSIS (Technique for Order Preference by Similarity to an Ideal Solution) method has been used for determining the distance to the ideal point, whereby the selected best alternative has the smallest distance to the best decision, and the largest distance to the worst decision [47]. The ARAS (Additive Ratio Assessment) method has indicated the best alternative that is closest to the optimal solution [48]. The EDAS (Evaluation Based on Distance from Average Solution) method has determined the best alternative as related to the distance from the average decision [49].

Applying various MCDM methods in solving the same problem frequently shows different assessment results. The Borda [50,51] and Copeland [52,53] methods can be used to identify the most significant alternatives computed by employing MCDM techniques.

Results of assessing the biggest Lithuanian cities in accordance with thematic areas have been obtained using MCDM methods, and are presented in Table 2. In line with different MCDM methods, priority lines differ slightly, and therefore the use of the Weighted average, Borda and Copeland methods assists in calculating the summarized priority lines.

Assessment of the results using the Weighted average, Borda and Copeland methods demonstrates that Vilnius and Panevėžys have collected the same number of ranks in the thematic area T4, and therefore take the first and second place in order of priority. To solve the above-introduced situation, the location of the priority line has been specified by averaging city significance in line with the findings obtained by performing MCDM assessment. Thus, Vilnius is ranked first (0.4636) and Panevėžys is ranked second (0.4325).

Analysis of the assessment results shows that the dominant city differs in each thematic area, and there is not a single city that has the highest ranks in all thematic areas; for example, the rank of Panevėžys in thematic area T3—Traffic safety and transport security is the highest, whereas their rank in thematic area T7—Promotion of alternative fuels and clean vehicles is the lowest.

To identify the city that has the highest rank pursuant to the established ranks of individual thematic areas, the ranks of the thematic areas of each city are summed. The city having the lowest calculated sum of ranks is given the highest rank and, in accordance with the same principle, the city with the highest calculated sum of ranks is given the lowest assessment rank. The obtained results are provided in Table 3.

The overall assessment of SUM disclosed that Vilnius city had the highest rank (sum of ranks was 18), and Šiauliai city had the lowest rank (sum of ranks was 30). Analysis of the sums of the ranks of the Kaunas and Klaipėda cities shows that the SUM of these cities were very similar, and therefore the mobility measures implemented in future are likely to change the situation in the general order of priorities.

**Table 2.** Assessment results applying MCDM methods.

| City | MCDM Methods | | | | | | | | Significance Average ($\tilde{\omega}$) | Location Considering Significance | Location Considering Average | Location as Stated in BORDA | Location as Stated in Copeland |
|---|---|---|---|---|---|---|---|---|---|---|---|---|---|
| | COPRAS | | TOPSIS | | ARAS | | EDAS | | | | | | |
| | Weight | Rank | Weight | Rank | Weight | Rank | Weight | Rank | | | | | |
| T1—Promotion of public transport | | | | | | | | | | | | | |
| Vilnius | 0.2188 | 2 | 0.6242 | 2 | 0.1588 | 2 | 0.7655 | 2 | 0.4418 | 2 | 2 | 2 | 2 |
| Kaunas | 0.1716 | 3 | 0.4723 | 3 | 0.1280 | 3 | 0.5233 | 3 | 0.3238 | 3 | 3 | 3 | 3 |
| Klaipėda | 0.2565 | 1 | 0.7322 | 1 | 0.1857 | 1 | 0.9654 | 1 | 0.5350 | 1 | 1 | 1 | 1 |
| Šiauliai | 0.0772 | 5 | 0.0806 | 5 | 0.0601 | 5 | 0.0016 | 5 | 0.0549 | 5 | 5 | 5 | 5 |
| Panevėžys | 0.0931 | 4 | 0.1403 | 4 | 0.0717 | 4 | 0.0905 | 4 | 0.0989 | 4 | 4 | 4 | 4 |
| T2—Non-motor vehicle integration | | | | | | | | | | | | | |
| Vilnius | 0.2873 | 1 | 0.5239 | 1 | 0.1885 | 1 | 0.7767 | 1 | 0.4441 | 1 | 1 | 1 | 1 |
| Kaunas | 0.1547 | 3 | 0.2997 | 3 | 0.1028 | 4 | 0.2694 | 3 | 0.2067 | 3 | 3 | 3 | 3 |
| Klaipėda | 0.1517 | 4 | 0.2841 | 4 | 0.1034 | 3 | 0.2269 | 4 | 0.1915 | 4 | 4 | 4 | 4 |
| Šiauliai | 0.1256 | 5 | 0.2612 | 5 | 0.0884 | 5 | 0.1148 | 5 | 0.1475 | 5 | 5 | 5 | 5 |
| Panevėžys | 0.2808 | 2 | 0.5210 | 2 | 0.1841 | 2 | 0.7285 | 2 | 0.4286 | 2 | 2 | 2 | 2 |
| T3—Traffic safety and transport security | | | | | | | | | | | | | |
| Vilnius | 0.1593 | 3 | 0.1876 | 3 | 0.1262 | 3 | 0.1789 | 3 | 0.1630 | 3 | 3 | 3 | 3 |
| Kaunas | 0.2030 | 2 | 0.3560 | 2 | 0.1540 | 2 | 0.3307 | 2 | 0.2609 | 2 | 2 | 2 | 2 |
| Klaipėda | 0.1317 | 5 | 0.1463 | 4 | 0.1064 | 5 | 0.0198 | 5 | 0.1011 | 5 | 5 | 5 | 5 |
| Šiauliai | 0.1363 | 4 | 0.1445 | 5 | 0.1066 | 4 | 0.0790 | 4 | 0.1166 | 4 | 4 | 4 | 4 |
| Panevėžys | 0.3697 | 1 | 0.6634 | 1 | 0.2289 | 1 | 0.8566 | 1 | 0.5297 | 1 | 1 | 1 | 1 |
| T4—Improvement to traffic organization and mobility management | | | | | | | | | | | | | |
| Vilnius | 0.2875 | 1 | 0.5653 | 2 | 0.1780 | 2 | 0.8236 | 1 | 0.4636 | 1 | 1-2 | 1-2 | 1-2 |
| Kaunas | 0.1226 | 5 | 0.1670 | 5 | 0.0669 | 5 | 0.0269 | 3 | 0.0959 | 5 | 5 | 5 | 5 |
| Klaipėda | 0.1629 | 4 | 0.3388 | 4 | 0.1204 | 4 | 0.2502 | 4 | 0.2181 | 4 | 4 | 4 | 4 |
| Šiauliai | 0.1706 | 3 | 0.4363 | 3 | 0.1477 | 3 | 0.2369 | 5 | 0.2479 | 3 | 3 | 3 | 3 |
| Panevėžys | 0.2563 | 2 | 0.5973 | 1 | 0.1988 | 1 | 0.6774 | 2 | 0.4325 | 2 | 1-2 | 1-2 | 1-2 |
| T5—City logistics | | | | | | | | | | | | | |
| Vilnius | 0.4474 | 5 | 0.1257 | 5 | 0.0354 | 5 | 0.000 | 5 | 0.0521 | 5 | 5 | 5 | 5 |
| Kaunas | 0.2226 | 2 | 0.5538 | 2 | 0.1421 | 2 | 0.4874 | 2 | 0.3515 | 2 | 2 | 2 | 2 |
| Klaipėda | 0.1092 | 3 | 0.2797 | 3 | 0.0775 | 3 | 0.2356 | 3 | 0.1755 | 3 | 3 | 3 | 3 |
| Šiauliai | 0.0507 | 4 | 0.1403 | 4 | 0.0379 | 4 | 0.0143 | 4 | 0.0608 | 4 | 4 | 4 | 4 |
| Panevėžys | 0.3807 | 1 | 0.6597 | 1 | 0.2279 | 1 | 1.0000 | 1 | 0.5671 | 1 | 1 | 1 | 1 |
| T6—Integration of people with special needs | | | | | | | | | | | | | |
| Vilnius | 0.1519 | 4 | 0.4559 | 4 | 0.1209 | 4 | 0.2666 | 4 | 0.2488 | 4 | 4 | 4 | 4 |
| Kaunas | 0.1425 | 5 | 0.2787 | 5 | 0.1144 | 5 | 0.0665 | 5 | 0.1505 | 5 | 5 | 5 | 5 |
| Klaipėda | 0.1550 | 3 | 0.4671 | 3 | 0.1236 | 3 | 0.3196 | 3 | 0.2663 | 3 | 3 | 3 | 3 |
| Šiauliai | 0.1675 | 2 | 0.5283 | 2 | 0.1337 | 2 | 0.5258 | 2 | 0.3388 | 2 | 2 | 2 | 2 |
| Panevėžys | 0.2094 | 1 | 0.9859 | 1 | 0.1665 | 1 | 1.0000 | 1 | 0.5905 | 1 | 1 | 1 | 1 |
| T7—Promotion of alternative fuels and clean vehicles | | | | | | | | | | | | | |
| Vilnius | 0.2107 | 1 | 0.6324 | 1 | 0.1502 | 1 | 0.9195 | 1 | 0.4782 | 1 | 1 | 1 | 1 |
| Kaunas | 0.1566 | 2 | 0.5218 | 2 | 0.1141 | 2 | 0.5226 | 2 | 0.3288 | 2 | 2 | 2 | 2 |
| Klaipėda | 0.1043 | 4 | 0.2700 | 5 | 0.0754 | 4 | 0.1518 | 4 | 0.1504 | 4 | 4 | 4 | 4 |
| Šiauliai | 0.1141 | 3 | 0.3654 | 3 | 0.0830 | 3 | 0.2037 | 3 | 0.1916 | 3 | 3 | 3 | 3 |
| Panevėžys | 0.0985 | 5 | 0.2746 | 4 | 0.0721 | 5 | 0.0722 | 5 | 0.1294 | 5 | 5 | 5 | 5 |
| T8—Assessment of demand for intelligent transport systems | | | | | | | | | | | | | |
| Vilnius | 0.2462 | 1 | 0.7837 | 1 | 0.1649 | 1 | 1.0000 | 1 | 0.5487 | 1 | 1 | 1 | 1 |
| Kaunas | 0.0914 | 3 | 0.3465 | 3 | 0.0617 | 3 | 0.3193 | 3 | 0.2047 | 3 | 3 | 3 | 3 |
| Klaipėda | 0.1774 | 2 | 0.4877 | 2 | 0.1192 | 2 | 0.6590 | 2 | 0.3608 | 2 | 2 | 2 | 2 |
| Šiauliai | 0.0414 | 4 | 0.2150 | 4 | 0.0277 | 4 | 0.0673 | 4 | 0.0879 | 4 | 4 | 4 | 4 |
| Panevėžys | 0.0296 | 5 | 0.1469 | 5 | 0.0198 | 5 | 0.0000 | 5 | 0.0491 | 5 | 5 | 5 | 5 |

**Table 3.** The overall assessment of SUM pursuant to the ranks of all thematic areas.

| City | T1 Rank | T2 Rank | T3 Rank | T4 Rank | T5 Rank | T6 Rank | T7 Rank | T8 Rank | Rank Sum | Final Rank |
|---|---|---|---|---|---|---|---|---|---|---|
| Vilnius | 2 | 1 | 3 | 1 | 5 | 4 | 1 | 1 | 18 | 1 |
| Kaunas | 3 | 3 | 2 | 5 | 2 | 5 | 2 | 3 | 25 | 3 |
| Klaipėda | 1 | 4 | 5 | 4 | 3 | 3 | 4 | 2 | 26 | 4 |
| Šiauliai | 5 | 5 | 4 | 3 | 4 | 2 | 3 | 4 | 30 | 5 |
| Panevėžys | 4 | 2 | 1 | 2 | 1 | 1 | 5 | 5 | 21 | 2 |

The impact of each thematic area on the overall assessment result is different, and thus using the significance of the thematic areas identified by experts helps with combining the individual MCDM methods so as to assess the overall development levels of SUM. In this case, the elements of the decision matrices are the values of the assessed thematic areas derived by applying individual MCDM methods. The maximum values obtained using the MCDM methods are the best results, and therefore all indicators of the thematic areas are maximized during the assessment process. Subsequently, the assessment of thematic areas using individual MCDM methods, and the overall assessment, are performed (see Table 4).

**Table 4.** The overall assessment of SUM pursuant to the weights of all thematic areas.

| City | COPRAS | | TOPSIS | | ARAS | | EDAS | | Significance Average ($\tilde{\omega}$) | Location Considering Average | Location as Stated in Borda | Location as Stated in Copeland |
|---|---|---|---|---|---|---|---|---|---|---|---|---|
| | Weight | Rank | Weight | Rank | Weight | Rank | Weight | Rank | | | | |
| Vilnius | 0.2456 | 1 | 0.5297 | 2 | 0.1836 | 1 | 0.7333 | 2 | 0.4231 | 2 | 2 | 2 |
| Kaunas | 0.1857 | 4 | 0.4159 | 4 | 0.1389 | 4 | 0.3298 | 4 | 0.2676 | 4 | 4 | 4 |
| Klaipėda | 0.1968 | 3 | 0.4651 | 3 | 0.1523 | 3 | 0.3683 | 3 | 0.2956 | 3 | 3 | 3 |
| Šiauliai | 0.1295 | 5 | 0.2043 | 5 | 0.1074 | 5 | 0.0145 | 5 | 0.1139 | 5 | 5 | 5 |
| Panevėžys | 0.2425 | 2 | 0.5609 | 1 | 0.1786 | 2 | 0.7713 | 1 | 0.4383 | 1 | 1 | 1 |

The analysis of the produced results disclosed that Vilnius and Panevėžys collected the same number of ranks, and shared the first and second places in order of priority. To solve the situation, the location of the priority line was specified according to the averages of the values obtained via MCDM assessment. Hence, Panevėžys was ranked first (0.4383), and Vilnius took the second position (0.4231).

The summary of the obtained findings demonstrates that equal overall results were calculated by all three of the Weighted average, Borda and Copeland methods. MCDM assessment showed the highest levels of SUM development in Panevėžys and Vilnius, rather than in other large cities. The values obtained in Vilnius and Panevėžys have been found to vary from those identified in Klaipėda city, which occupied the next place in terms of priority by more than 40%. This shows a gap in the implementation of the sustainable urban mobility measures between the cities.

## 6. Discussion and Conclusions

A comparison of two types of assessment has shown varying results in term of priority. Differences in the findings demonstrate that ranking significance does not consider the actual level of implementation of the urban mobility measure within the MCDM assessment process. Ranking only states the fact that the appropriate significance of a thematic area of a certain city is the highest compared to other cities, but does not point out the significance of that thematic area in the general transport system. The results of both types of assessment lead to the conclusion that, for overall ranking by aggregating the individual ranks of thematic areas, a city that enacts more mobility measures compared to other cities can be identified. Further, the determination of the significance of the thematic areas established by the experts, and the combination of thematic area values determined via MCDM methods, both assist in distinguishing the cities that are implementing more higher-quality (more significant) mobility measures.

The results of the undertaken assessment show that it is appropriate to use the ranking method in determining the cities occupying the leading positions with regard to individual thematic areas. However, the numerical significance of thematic areas needs to be considered when assessing the overall level of SUM development.

Table 4 shows that the acquired average significance of the cities can be used as an index for SUM development, showing the relative progress of a city in implementing sustainable mobility measures compared to other cities of the same type. For comparing these indexes with each other, the possibility of determining the differences in the relative effectiveness of the SUM development levels in individual cities is suggested. For instance, comparing Panevėžys city (0.4383) with Šiauliai city (0.1139) demonstrates that the effectiveness of the SUM development level in Panevėžys is 3.85-fold higher than that in Šiauliai. Such a comparison is used in practice when planning the development of the common transport system, providing measures to be implemented and calculating clear and comparable indexes for the decisions taken.

The findings do not assume that the development of SUM should mainly focus only on the most significant mobility measure or thematic area, because successful SUM planning is determined by the integrated implementation of mobility measures. However, it should be noted that the smooth implementation of both low and greater mobility measures will result in a higher overall level of developing SUM, and better conditions for the society.

The significance levels of the thematic areas and mobility measures identified during expert consultation can constitute a good guide in assessing the mobility measures selected for implementation. In order to achieve maximum results with the least time and money, it is recommended to implement the mobility measures which, firstly, are appropriate for the local context, and secondly, have a greater relevance to and impact on the mobility of the society. MCDM assessment assists in estimating the effect that mobility measures will have on the transport system, comparing between similar types of cities.

A combination of different MCDM methods and a summary of the obtained results both allow for more accurate results in assessing the current/planned effectiveness of transport systems. This instrument is particularly useful for 'beginner' cities, which are starting to step up their implementation of sustainable mobility measures and are looking for practical examples in similar cities. The suggested model demonstrates the effectiveness of the transport system, considering the aspect of consumer satisfaction rather than cost-effectiveness (e.g., using cost–benefit analysis), i.e., the choice and implementation of mobility measures is assessed through the need and impact of infrastructure and services.

The proposed model is not applied to all European cities (e.g., metropolitan areas), and therefore leaves room for further work and research on the list of mobility measures, and the identification of their relevance, for example, to metropolitan areas.

**Author Contributions:** Conceptualization, J.D. and M.B.; methodology, J.A.; investigation, J.D.; data curation, J.D. and J.A.; writing—original draft preparation, J.D.; writing—review and editing, J.A.; visualization, J.D.; supervision, M.B. All authors have read and agreed to the published version of the manuscript.

**Funding:** This research was funded by Vilnius Gediminas Technical University.

**Conflicts of Interest:** The authors declare no conflict of interest.

## References

1. European Commission (EC). White Paper. Roadmap to a Single European Transport Area—Towards a competitive and resource efficient transport system. 2011. Available online: http://eur-lex.europa.eu/legal-content/EN/ALL/?uri=CELEX:52011DC0144 (accessed on 22 February 2020).
2. European Commission (EC). Communication from the Commission to the European Parliament, the Council, the European Economic and Social Committee and the Committee of the Regions Together towards competitive and Resource—Efficient Urban Mobility. 2013. Available online: https://eur-lex.europa.eu/legal-content/EN/TXT/?qid=1585991520554&uri=CELEX:52013DC0913 (accessed on 22 February 2020).
3. Rupprecht Consult 2019. Guidelines for Developing and Implementing a Sustainable Urban Mobility Plan, Second Edition. Available online: https://www.eltis.org/sites/default/files/sump-guidelines-2019_mediumres.pdf (accessed on 11 May 2020).
4. Ministry of Transport and Communications of the Republic of Lithuania (MoTC). National Guidelines of Preparation of Sustainable Urban Mobility Plans. 2015. Available online: https://e-seimas.lrs.lt/portal/legalAct/lt/TAD/a1c919e0c9cc11e4bc22872d979254dd?jfwid=-l5uh8xdz3 (accessed on 6 February 2020).
5. May, A.D.; Kelly, C.; Shepherd, S.; Jopson, A. An option generation tool for potential urban transport policy packages. *Transp. Policy* **2012**, *20*, 162–173. [CrossRef]
6. Sundberg, R. CIVITAS SUPs-Up Manual on the Integration of Measures and Measure Packages in a SUMP for Advanced Cities. 2018. Available online: https://sumps-up.eu/publications-and-reports (accessed on 7 May 2020).
7. Sundberg, R. CIVITAS SUPs-Up Manual on the Integration of Measures and Measure Packages in a SUMP for Beginner Cities. 2018. Available online: https://sumps-up.eu/publications-and-reports (accessed on 7 May 2020).

8. Sundberg, R. CIVITAS SUPs-Up Manual on the Integration of Measures and Measure Packages in a SUMP for Intermediate Cities. 2018. Available online: https://sumps-up.eu/publications-and-reports (accessed on 7 May 2020).
9. Shergold, I.; Parkhurst, G. Demonstrating Economic Benefit from Sustainable Mobility Choices—The EVIDENCE Project. 2017. Available online: https://ec.europa.eu/energy/intelligent/projects/sites/iee-projects/files/projects/documents/evidence_publishable-report_final_2017.pdf (accessed on 7 May 2020).
10. The Urban Transport Roadmap (UTR). Study on European Urban Transport Roadmap 2030. 2016. Available online: http://www.urban-transport-roadmaps.eu (accessed on 7 May 2020).
11. KonSULT. Knowledgebase on Sustainable Urban Land Use and Transport. 2016. Available online: http://www.konsult.leeds.ac.uk/ (accessed on 6 February 2020).
12. May, A.D. CH4LLENGE Manual on Measure Selection: Selecting the most Effective Packages of Measures for Sustainable Urban Mobility Plans. 2016. Available online: http://www.sump-challenges.eu/ (accessed on 11 May 2020).
13. Chakhtoura, C.; Pojani, D. Indicator-based evaluation of sustainable transport plans: A framework for Paris and other large cities. *Transp. Policy* **2016**, *50*, 15–28. [CrossRef]
14. Nocera, S.; Tonin, S.; Cavallaro, F. Carbon estimation and urban mobility plans: Opportunities in a context of austerity. *Res. Transp. Econ.* **2015**, *51*, 71–82. [CrossRef]
15. Van Audenhove, F.J.; Koriichuk, O.; Dauby, L.; Pourbarx, J. The Future of Urban Mobility 2.0: Imperatives to Shape Extended Mobility Ecosystems of Tomorrow. 2014. Available online: http://www.uitp.org/sites/default/files/members/140124%20Arthur%20D.%20Little%20&%20UITP_Future%20of%20Urban%20Mobility%202%200_Full%20study.pdf (accessed on 7 May 2020).
16. Shiau, T.A.; Huang, M.W.; Lin, W.Y. Developing an indicator system for measuring Taiwan's transport sustainability. *Int. J. Sustain. Transp.* **2013**, *9*, 81–92. [CrossRef]
17. Haghshenas, H.; Vaziri, M.; Gholamialam, A. Evaluation of sustainable policy in urban transportation using system dynamics and world cities data: A case study in Isfahan. *Cities* **2012**, *45*, 104–115. [CrossRef]
18. MaxExplorer Tool. 2006. Available online: http://www.epomm.eu/index.php?id=2745 (accessed on 11 May 2020).
19. Arsenio, E.; Martens, K.; Ciommo, F. Sustainable urban mobility plans: Bridging climate change and equity targets? *Res. Transp. Econ.* **2016**, *55*, 30–39. [CrossRef]
20. Reisi, M.; Aye, L.; Rajabifard, A.; Ngo, T. Transport sustainability index: Melbourne case study. *Ecol. Indic.* **2014**, *43*, 288–296. [CrossRef]
21. Lima, J.P.; Lima, R.S.; Silva, A.N.R. Evaluation and selection of alternatives for the promotion of sustainable urban mobility. *Procedia Soc. Behav. Sci.* **2014**, *162*, 408–418. [CrossRef]
22. Saisana, M. Weighting methods. In Proceedings of the Seminar on Composite Indicators: From Theory to Practice, Ispra, Italy, 13–14 January 2011.
23. Mancini, M.T. Urban Planning Based on Scenarios of Sustainable Mobility. Master's Thesis, São Carlos School of Engineering, University of São Paulo, São Paulo, Brazil, 2011.
24. Silva, A.N.R.; Costa, M.S.; Ramos, R.A.R. Development and Application of I_SUM: An Index of Sustainable Urban Mobility. Transportation Research Board Annual Meeting. 2010. Available online: http://citeseerx.ist.psu.edu/viewdoc/download?doi=10.1.1.1008.9281&rep=rep1&type=pdf (accessed on 29 April 2020).
25. Costa, M.S. An Index of Sustainable Urban Mobility. Ph.D. Thesis, São Carlos School of Engineering, University of São Paulo, São Paulo, Brazil, 2008.
26. Zhou, P.; Ang, B.; Poh, K. A mathematical programming approach to constructing composite indicators. *Ecol. Econ.* **2007**, *62*, 291–297. [CrossRef]
27. Burinskienė, M.; Gaučė, K.; Damidavičius, J. Successful sustainable mobility measures selection. In Proceedings of the 10th International Conference "Environmental Engineering", Vilnius, Lithuania, 27–28 April 2017. [CrossRef]
28. Shiau, T.A.; Liu, J.S. Developing an indicator system for local governments to evaluate transport sustainability strategies. *Ecol. Indic.* **2013**, *34*, 361–371. [CrossRef]
29. Karagiannakidis, D.; Sdoukopoulos, A.; Gavanas, N.; Pitsiava-Latinopoulou, M. Sustainable urban mobility indicators for medium-sized cities. In Proceedings of the case of Serres, Greece, 2nd Conference on Sustainable Urban Mobility, Valos, Greece, 5–6 May 2012.

30. Malasek, J. A set of tools for making urban transport more sustainable. *Transp. Res. Procedia* **2016**, *14*, 876–885. [CrossRef]
31. Litman, T. Well Measured. Developing Indicators for Sustainable and Liveable Transport Planning. 2016. Available online: http://www.vtpi.org/wellmeas.pdf (accessed on 22 February 2020).
32. Rupprecht Consult. CH4LLENGE—Addressing Key Challenges of Sustainable Urban Mobility Planning. 2016. Available online: http://www.sump-challenges.eu (accessed on 11 May 2020).
33. Litman, T. Sustainable Transportation Indicators—A Recommended Research Program for Developing Sustainable Transportation Indicators and Data. In Proceedings of the 88th Annual Meeting of the Transportation Research Board, Washington, DC, USA, 11–15 January 2009.
34. Greco, S.; Matarazzo, B.; Slowiński, R. Rough sets theory for multicriteria decision analysis. *Eur. J. Oper. Res.* **2001**, *129*, 1–47. [CrossRef]
35. Kendall, M. *Rank Correlation Methods;* Published by Griffin: London, UK, 1970.
36. Kendall, M.; Gibbons, J.D. *Rank Correlation Methods*, 5th ed.; Oxford University Press: Oxford, UK, 1990; 272p.
37. Zavadskas, E.K.; Peldschus, F.; Kaklauskas, A. *Multiple Criteria Evaluation of Projects in Construction*; Technika: Vilnius, Lithuania, 1994.
38. Zavadskas, E.K.; Vilutiene, T. A multiple criteria evaluation of multi-family apartment block's maintenance contractors: I—Model for maintenance contractor evaluation and the determination of its selection criteria. *Build. Environ.* **2006**, *41*, 621–632. [CrossRef]
39. Jakimavicius, M.; Burinskiene, M.; Gusaroviene, M.; Podviezko, A. Assessing multiple criteria for rapid bus routes in the public transport system in Vilnius. *Public Transp.* **2016**, *8*, 365–385. [CrossRef]
40. Parezanovic, T.; Bojkovic, N.; Petrovic, M.; Pejčic-Tarle, S. Evaluation of Sustainable Mobility Measures Using Fuzzy COPRAS Method. *J. Sustain. Bus. Manag. Solut. Emerg. Econ.* **2016**, *21*, 53–62. [CrossRef]
41. Hickman, R.; Saxena, S.; Banister, D.; Ashiru, O. Examing transport future with scenario analysis and MCA. *Transp. Res.* **2012**, *46*, 560–575. [CrossRef]
42. Podvezko, V.; Sivilevičius, H. The use of AHP and rank correlation methods for determining the significance of the interaction between the elements of a transport system having a strong influence on traffic safety. *Transport* **2013**, *28*, 389–403. [CrossRef]
43. Barauskas, A.; Mateckis, K.J.; Palevičius, V.; Antuchevičiene, J. Ranking conceptual locations for a park-and-ride parking lot using EDAS method. *Građevinar* **2018**, *70*, 975–983. [CrossRef]
44. Ginevičius, R.; Podvezko, V.; Podviezko, A. Evaluation of Isolated Socio-Economical Processes by a Multi-Criteria Decision Aid Method ESP. In Proceedings of the 7th International Scientific Conference Business and Management', Vilnius, Lithuania, 10–11 May 2012.
45. Palevičius, V.; Grigonis, V.; Podviezko, A.; Barauskaite, G. Developmental analysis of park-and-ride facilities in Vilnius. *Promet Traffic Transp.* **2016**, *28*, 165–178. [CrossRef]
46. Zavadskas, E.K.; Kaklauskas, A. *Pastatų Sistemotechninis Įvertinimas;* Technika: Vilnius, Lithuania, 1996; 275p.
47. Hwang, C.L.; Yoon, K. *Multiple Attribute Decision-Making Methods and Applications;* A State of the Art Survey; Springer: Berlin, Germany, 1981.
48. Zavadskas, E.K.; Turskis, Z. A new additive ratio assessment (ARAS) method in multicriteria decision-making. *Technol. Econ. Dev. Econ.* **2010**, *16*, 159–172. [CrossRef]
49. Keshavarz Ghorabaee, M.; Zavadskas, E.K.; Olfat, L.; Turskis, Z. Multi-criteria inventory classification using a new method of Evaluation based on Distance from Average Solution (EDAS). *Informatica* **2015**, *26*, 435–451. [CrossRef]
50. Borda, J.C. *Memoire sur les Elections au Scrutiny, Histoire de l'Academie Royale des Sciences;* Paris, France, 1781; Available online: http://asklepios.chez.com/XIX/borda.htm (accessed on 7 May 2020).
51. McLean, I. The Borda and Condorcet principles: Three medieval applications. *Soc. Choice Welf.* **1990**, *2*, 99–108. [CrossRef]
52. Erlandson, R. System Evaluation Methodologies: Combined Multi-dimensional Scaling and Ordering Techniques. *IEEE Trans. Syst. Man Cybern.* **1978**, *6*, 421–432. [CrossRef]
53. Fishburn, P. A Comparative Analyses of Group Decision Methods. *Behav. Sci.* **1971**, *16*, 538–544. [CrossRef]

*Article*

# Policies for Reducing Car Traffic and Their Problematisation. Lessons from the Mobility Strategies of British, Dutch, German and Swedish Cities

**Tom Rye [1,2,*] and Robert Hrelja [3,4]**

1 Faculty of Logistics, Molde University College, Britvegen 2, 6410 Molde, Norway
2 Urban Planning Institute of the Republic of Slovenia, Trnovski pristan 2, 1000 Ljubljana, Slovenia
3 Department of Urban Studies, Malmo University, 20506 Malmo, Sweden; robert.hrelja@mau.se
4 K2—The Swedish Knowledge Centre for Public Transport, Bruksgatan 8, 223 81 Lund, Sweden
* Correspondence: tom.rye@himolde.no

Received: 28 August 2020; Accepted: 28 September 2020; Published: 3 October 2020

**Abstract:** The objective of the paper is to explore whether particular problematisations of cars and car use lead to sets of solutions that may not deal with all problems associated with car use, and whether this leads to any internal conflicts within the chosen policies. The paper is based on a review of local transport policy documents from 13 cities in four countries using the lens of policy problematisation as an analytical framework. Some critiques of policy problematisation are discussed in the paper but it is nonetheless shown to be helpful for this analysis. The paper finds that the problems most typically highlighted in the strategies reviewed are poor accessibility (as a "bad" in itself, but also because it is seen to compromise economic growth); the negative impacts of traffic on liveability of the central part of the city and therefore its ability to attract inhabitants, especially those needed to support a knowledge economy; local air and noise pollution; and road safety. The resulting visions are for urban areas less dominated by private cars, with more green and public space, in order to maximise accessibility and liveability to attract economic development; and most cities also seek to reduce car travel as a proportion of trips. However, in many cities this vision covers mainly the central city, with car use set to remain dominant in outer cities and for regional trips. In almost all cities, only one measure, parking management, is proposed as a means of cutting car use. The differing sets of measures envisaged for outer areas of cities threatens to undermine those envisaged for more central cities.

**Keywords:** policy; problematisation; local transport; mobility plan; Sweden; Great Britain; Netherlands; Germany

---

## 1. Introduction

Urban transport, since World War Two if not before [1–5], has been planned with a focus on the private car, and in most cities and countries this focus is still the norm in planning practices. On the other hand, there is an increasing policy imperative to make cities and urban regions more sustainable in transport terms, with less motor vehicle use—for example, the EU's policy goal to make urban transport fossil fuel free by 2030. This can often lead to conflicts between automobility (a car-based norm) and sustainability goals in urban transport policy.

This conflict arises from the problems caused by automobility and sustainability goals, including, among other things, high capacity road systems that induce traffic, or traffic congestion, periods of high particulate levels, reduced physical activity, poor road safety, carbon dioxide emissions, high noise levels along major roads and so on. However, it is not obvious which of these problems is the most important to solve, nor how to solve them.

This begs the following questions, which are the basis of this paper:

- How are the problems caused by cars framed by politicians and planners when developing their plans for a more sustainable transport system in their cities?
- Does this have an influence on the policies and interventions included in those plans?

The analytical assumption is that the way cars are intentionally framed as a problem in the transport policy-making process will have a significant influence on the choice of objectives as well as measures selected to achieve those objectives; and if problems are deliberately or accidentally left out of this problematisation process, this may (a) lead to conflicts within the resulting policy, and (b) lead to path-dependencies in planning that continue to favour car-based mobility futures. In this paper we apply, following Bacchi [6,7], the lens of policy problematisation to local transport policy making in urban municipalities in four European countries, through a review of their transport policy documents. This theoretical approach draws attention to how policy is created in discourse. Theoretically, it is assumed that the framing of policy problems is a necessary part of the policy-making process. It also assumes, as indicated above, that a specific problematisation tends to lead to certain consequences resulting from how the "problem" is constructed and demarcated. For example, if excessive car traffic leading to congestion in city centres is defined as a problem resulting from a lack of public transport as a suitable alternative to private cars, then the most likely policy response will be to try to improve the quality of public transport, instead of (for example) implementing car restrictive measures on car traffic. The review of the transport policy documents is informed by this theoretical approach and, in line of it, identifies the dominant representations of problems, the assumptions on which these descriptions of dominant problems are based and the measures that are presented as appropriate to implement to solve the problem in question.

The objective of the paper is to explore whether particular problematisations of cars and car use lead to particular sets of solutions that may not deal with all the associated problems, and whether this leads to any internal conflicts within the chosen policies, and/or "silences" regarding problems that are known, but not acknowledged in the policies. The analysis focuses particularly on the role of the private car in cities' mobility futures and the measures selected to realise these futures; this focus is because the private car remains the most significant source of $CO_2$ emissions and cause of congestion in urban surface transport.

We further hypothesise that:

1. There is a particular definition of the desired mobility future in cities' transport policy documents.
2. This leads to a particular choice of measures dealing with car use in these cities.
3. This generates some possible conflicts between the measures selected (or not selected) and the mobility futures imagined.
4. The policy problematisations observed in the documents result in silences of importance for the transition towards sustainable transport systems—in terms of CO2 emissions and overall km travelled—that partly explain how path dependencies reported in previous research (e.g., [8,9]) are produced.

The rest of the paper is structured as follows. The literature review covers, from a scientific perspective, the key transport problems that cities and urban regions face today. It then summarises literature on policy problematisation, which provides the basis for the analytical framework used in the paper. It also caveats the analytical framework, in particular discussing alternative views of policy problematisation. It further covers relevant literature on path dependence in planning and the relevance of this to our study. The paper then explains the methodology used and the cities selected for the study (as well as the (deliberate) limitations of this selection), before presenting the empirical findings of the document review and relating these to the analytical framework. Finally, the paper draws a number of conclusions related to our initial objectives and hypotheses, and these are then used to provide some policy recommendations on how to plan for decreased car use.

## 2. Literature Review Including Analytical Framework

There are many potential measures to be used [10] when creating more sustainable road transport systems and when reducing urban car travel (see [11] for an overview). Measures may either be [12] intended to "pull" passengers to other modes of transport (by making walking, biking and public transport an attractive alternative), or to "push" passengers to other modes of transport (by making the car a less attractive option). Measures such as restricting speeds, reducing road capacity and the management of parking spaces fall into this latter category. Research has also shown which measures that have typically been implemented by those few cities have reduced car use as a proportion of trips (see for example [13–15] for European examples). There is also knowledge of some of the factors that make it possible to implement such measures, such as reorganisation of local administrations [16], the important role of green policy entrepreneurs in local authorities, and research giving concrete policy recommendations for implementing sustainable transport goals, for example about [13] how controversial policies should be implemented in stages. Thus, there are several studies that provide concrete advice on what could be done and how to create change in what has traditionally been a very car-centric planning praxis. At the same time, there is a great deal of research that shows how difficult it is to reduce car use in urban areas. The sustainable transformation of cities' transport systems is often constrained by barriers such as rebound effects, conflicting visions at different levels, lack of consensus among stakeholders, public objections and path dependencies [17,18]. The key issue arising from this short review is how to accelerate the transition towards sustainable transport. This in turn presupposes knowledge of why car-centric planning is produced and reproduced by discourses, established practices, routines and methods applied in planning.

At the same time, researchers have argued that the orientation of transport research means that it rarely asks important questions about how to how to accelerate the transition towards sustainable transport. Marsden and Reardon identify some important gaps in the transportation policy literature that are the consequence of what they call a dominant "techno-rational" approach to studying transportation policy [19]. This dominant approach focuses, according to Marsden and Reardon, on the means and tools of policy making, and little attention is generally paid to the goals and settings of policy making: [19] (p. 245) "The 'policy' literature is therefore currently drawn to answering questions relating to what is, and making that work more effectively, than on critiquing the assumptions of the current status quo, and arguing for what ought to be, or what could be." Their analysis shows that there is a need for a more reflective discussion on how policy is shaped in order to better understand the opportunities for change, a question that, in this paper, relates to ways to reduce motor vehicle traffic.

To meet this research need, we have adopted a theoretical approach to policy that is inspired by research in a critical policy studies tradition that understands policy [20] (p. 1) "in terms of the interests, values and normative assumptions ... that shape and inform [policy] processes". In analysing how car travel might be reduced in urban transport planning, we are analytically interested in discussing which values are prioritised and how values influence which policy instruments that are seen as appropriate in the local contexts studied.

There are, following [21], three interrelated factors that may produce path dependencies in transport policy and planning: *institutional factors*, relating to practices, routines and methods applied by key organisations; *technical factors*, relating to the momentum resulting from fixed infrastructure serving societal functions; and *discursive factors*, relating to assumptions, justifications or beliefs, that apply within an organisation and shape its practices. Analytically, this paper discusses how policy is created in discourse, allowing a focus on the practices by which conceptions of knowledge and meaning are produced and reproduced in policy practices, with the consequent production of [22] dominant modes of thought and behaviours. There are several theoretical approaches that share this interest, for example Rein and Schön's research into frame-critical policy analysis in which frames are defined as [23] (p. 89) "generic narratives ... that tell, within a given issue terrain, what needs fixing and how it might be fixed". Transport research building on Rein and Schön, for example [6] Richardson, Isaksson and Gullberg, illustrates how car-based automobility frames survive and continue to support car-based mobility futures even when seemingly radical policies are implemented (exemplified by these authors by the case of congestion charges in the Swedish capital of Stockholm).

Another theoretical approach is offered by [6,7] Bacchi and is used in this paper. Our theoretical starting point, following Bacchi, is that policies shape "problems", and that local politicians and transport planners are active in creating policy problems rather than reacting to problems "out there" in society. Such policy problematisations involve a "diagnostic" aspect that prescribes solutions to socially constructed problems. As such, a policy represents a particular way of understanding how, in this case, car use as a particular policy issue is created and used to mobilise decisions and implementation. However, problematisations define not just a problem, but also what is not a problem. Alternative problem formulations are eliminated by demarcation and ignored. In summary, the issue of car travel reduction is understood as parts of "problem representations" that create particular ways of understanding car traffic as a policy problem, which in turn influence the measures seen as appropriate or inappropriate in the local contexts studied here.

This theoretical approach has recently been applied to transport policy, also in a Swedish context by [24] Hrelja and [25] Rhenlund. The results from this research indicate that policy changes that may result in car travel reduction are indeed underway in many Swedish cities. At the heart of these changes are narratives about city development in which cities understand the "attractive city" as one where cars are defined as a problem to be addressed. However, the dominant policy problematisation also produce several "blind spots" of importance for the development of car use, which are discussed later in this paper.

The strength of Bacchi's approach is that it both provides a theoretical framework that can be used to analyse the "interests and normative assumptions that shape and policy processes", and that it provides an accompanying methodology. Bacchi suggests studying policy by posing a number of analytically motivated questions to interrogate the empirical material. Inspired by [6,7] Bacchi, in this study, the case cities' transport policy documents were analysed based on the following questions:

- What are the problems that are highlighted? Which problems are most frequently highlighted? Which figure, but are less important? Are there any cities or countries that have particularly unique problems identified? Are there problems that are ignored by the majority of or all cities? Do all cities actually engage in problematisation in these documents?
- Mobility futures imagined—commonalities and differences? What are the mobility futures imagined and how does car use fit into these? Of those cities that imagine a mobility future with an explicit recognition of less car use, do they have anything in common?
- Does the dominant policy problematisation produce any silences, conflicts or ambiguities?
- What measures are seen as appropriate or inappropriate? Are there any major differences in measures proposed linked to differences in policy problematisation?
- What consequences (for mobility, the city or its people and the environment) are produced by the chosen representation of the problem?

The theoretical approach adopted thus implies a specific relationship between policy problematisations and measures, in that the former is supposed to result in measures being understood as appropriate or inappropriate for dealing with the problem of car use. However, the relationship between problem and solutions is in reality often more complicated than this. Other research (often based on [26] Kingdon's Multiple Streams Approach (MSA)) recognises the importance of problematisation but also suggests that other factors influence how policies and related measures are developed. For example, it points out how measures as solutions may exist independently and before problems have been defined, and that there are solutions "floating" and waiting for a problem, which implies that available or measures may influence how policy problems (and related solutions) are constructed—yet not all solutions will ultimately be implemented, due to factors such as their relationship to existing values, and their technical and financial feasibility. The MSA also brings in concepts such as [27] the political context, the policy entrepreneur, and the window of policy timing.

Another potential problem of the theoretical approach based on Bacchi is the difficulty of linking discursive meaning to effects, that is, actual decisions about implementation of measures in practice. We have only examined which measures appear appropriate or inappropriate within the policy problematisations we identify and based only on a document analysis. This means that the results can only indicate potential effects of dominating policy problematisations and with this form of research we cannot find out whether and why the measures set out in the policy documents reviewed are actually implemented. It is clear from authors such as [28] that public attitudes play a key role in shaping the content of Sustainable Urban Mobility Plans (SUMPs), and the use of public participation activities to shape the documents that we have reviewed is referred to in all of them, but the nature and effects of these activities are not something that we examine directly in this paper, since it is a document analysis.

Whilst we recognise the value of the MSA approach, in this paper our focus is on an international comparison of what cities say that they will do, and why, in their written transport policy. For this, Bacchi's framework is more suitable and we draw some useful conclusions based on this approach. However, further, more in-depth research in a subset of our cities could usefully deploy the MSA to understand why the policies identified were chosen, and why some are implemented and others not.

## 3. Methodology

As highlighted earlier, the empirical work is based on a document analysis, a recognised qualitative technique [29–31]. Due to the limitations of the project resources, three cities were selected from each of England, Netherlands and Germany, and four from Sweden, making 13 in total whose transport strategy documents were analysed. Each city was in effect a case study but the data from each case study were limited to the document(s) analysed.

In a seminal piece of work on case study selection, Flyvberg [32] identified several strategies for case study selection, as follows:

A. Random selection, to avoid systematic biases in the sample.
B. Information oriented selection, to maximise the utility of information from small samples and single cases.
C. Extreme/deviant cases, to obtain information on unusual cases, which can be especially problematic or especially good.
D. Maximum variation cases, to obtain information about the significance of various circumstances for case process and outcome (cases that are very different on one dimension: size, form of organisation).
E. Critical cases, to achieve information that permits logical deductions of the type, "If this is (not) valid for this case, then it applies to all (no) cases".

For this study, we used strategy B, since there was necessarily a relatively small sample size. The sampling was based on pre-selected criteria, listed below. Without reading all the transport strategies of all similar cities in all four countries it is not possible to be certain that our chosen strategies are entirely representative of the "average" transport strategy in the country. However, beyond the criteria below, nothing indicated to us that our selected strategies were atypical of those in the respective country. The selection of case cities was based on the ambition to choose:

- Both those where trends in travel patterns have become more sustainable and those that have followed the national pattern.
- Both "leading cities", in the sense of being seen in by peers in their country as leading within the field of sustainable transport (Nottingham, Groningen, Lund and Lindau), and more average cities.
- At least one from each country with a specific objective in its strategy to reduce car traffic.

- Only smaller cities of between 40,000 and 300,000 people without (significant) light rail or tram networks, as the comparison of smaller with larger and especially capital cities with highly developed public transport networks would not be valid (other authors such as [33] have compared transport strategies from capital cities, although using a different methodology).

Further discussion of our city sample is provided in Section 4, below.

The selection strategy was thus one of choosing cities that are different on dimensions such as size and ambitions to obtain information about the significance of policy problematisations for the choice of measures that deal with the problems associated with car use.

In each city, the urban transport plan or equivalent was reviewed, along with the strategy/policy section of the land use plan, and other documents such as parking strategies if these were available (see Table 1 for full details of the documents analysed for each city).

**Table 1.** Cities and plans analysed.

| City | Transport Plan or Strategy | Spatial Plan |
|---|---|---|
| Aachen DE | Mobility Vision 2050 (2014) [34].<br>Mobility Strategy 2030 (*Verkehrsentwicklungsplan* (VEP)), 2015 [35]. | Spatial plan policy section (*Flachennutzplan*), 2014 [36]. |
| Bath UK | Getting Around Bath–a Transport Strategy for Bath, (2014) [37].<br>Draft joint, regional, Local Transport Plan (LTP) number 4 (2019) [38]. | Core Strategy of the Local Plan (2014) [39].<br>The Placemaking Strategy (2017) [40]. |
| Darlington UK | Third Local Transport Plan, (2011) [41]. | *Core Strategy* of the *Local Plan*–(2018) [42]. |
| Eindhoven NL | Mobility Vision 2040, *(Eindhoven op Weg)* (2013) [43].<br>New parking norms (*Nota Parkeernormen*) (2016) [44]. | The Land Use Strategy for housing (*Beleidsnota Prioriteiten bouwlocaties*), (2015)–it was not possible to find a policy section for the city's land use plan as a whole [45]. |
| Eskilstuna SE | Transport Strategy (*Strategidel*) 2012 [46]. | Comprehensive Plan (*Översiktsplan*), 2013 [47]. |
| Groningen NL | Mobility Strategy (*Nota Mobiliteit*) 2007-2010 (2007) [48].<br>Multi-year Programme for Transport and Traffic 2018-2021 (*Meerjarenprogramma Verkeer 2018–2021*) (2017) [49].<br>Parking Strategy (*Ruimte voor de Straat*) (2018) [50]. | Spatial Planning Strategy (*Nota Grondbeleid*) (2017) [51]. |
| Herrenberg DE | Integrated Mobility Development Plan (2019) [52]. | Various land use projects found on the city's website, accessed 23 October 2019 https://www.herrenberg.de/BayWa. |
| Jönköping SE | Transport Strategy (*Kommunikationsstrategi Åtgärder För Ett Hållbart Trafiksystem*) 2012 [53] | Comprehensive Plan (*Översiktsplan*), 2015 [54]. |
| Lindau DE | Climate Friendly Mobility Concept for Lindau (*Klimafreundliches Lindauer Mobilitätskonzept*) (2017) [55]. | Local land use zoning plan and supporting documents (*Flächennutzplan)*, (2011, modified 2013) [56].<br>Integrated City Development Concept (ICDC), *(Integriertes Stadtentwicklungskonzept)* (2015) [57] |
| Lund SE | Transport Strategy (*Trafikstrategi*) 2014 [58]. | Comprehensive Plan (*Översiktsplan*), 2017 [59]. |
| Malmö SE | Transport Strategy (*Trafikstrategi*) 2016 [60]. | Comprehensive Plan (*Översiktsplan*), 2018 [61] |
| Nottingham UK | Local Transport Plan 2016 [62]. | Local Plan 2017 [63]. |
| Tilburg NL | Mobility Strategy Tilburg, Together Towards Smart Sustainable Mobility in 2040 *(Mobiliteitsaanpak Tilburg Samen op weg naar 2040-Tilburg slim en duurzaam op weg)* (2017) [64].<br>Further development of the above *(Uitwerking Mobiliteitsaanpak Tilburg)* (2017) [65]. | Land Use Vision *(Omgevingsvisie Tilburg)*, 2015 [66] |

The plans were read in their original language. The analysis of the plans was performed stepwise including superficial examination of all plans, thorough examination, and interpretation – a qualitative research method [30,67,68]. All plans were first skimmed through and passages of text about transport, and especially about car traffic, was identified. Then a more thorough examination and re-reading of the plans was done. This step meant that the plans were read several times and key phrases were marked. English language summaries for each city were then produced relating the content of the policy documents to the analytical questions introduced earlier (Section 2). Recurrent regularities in the empirical material or themes and hierarchies of themes were also outlined in this way [67]. In order to reduce cognitive bias, the summaries were then reviewed by both research team members before being finalised [68]. This investigator or analyst triangulation provided an important check on interpretive bias in a situation where two researchers analysed the plans.

Any silences, conflicts or ambiguities produced by the dominating policy problematisation are judgements on our part as to what is actually a problem but is not seen as a problem in the plans. Silences were identified by reading the plans and by identifying text parts not associated with a theme or a policy problematisation.

In addition, the numbers of cities whose documents featured, for example, an explicit policy problematisation; or an explicit recognition that car use must be reduced; or clearly defined, specified and costed measures to achieve strategy objectives, were also identified.

The documents reviewed are of course produced within different national legal and institutional contexts. In order to maximise their comparability, as far as possible, strategic transport policy documents were reviewed containing a vision for future mobility, often a problem analysis, a set of objectives, often a set of targets (for example for mode share or improved road safety) and a set of measures to be implemented to achieve the objectives—although the level of detail to which the measures were specified varied quite widely across the documents reviewed, and time horizons varied from 10 to 30 years. The documents and their statutory basis are below (see also Table 1):

- England—Local Transport Plans, which were formerly statutory but are now optional, based around core objectives and measures to achieve these.
- Germany—Traffic Development Plan (*Verkehrsentwicklungsplan*), statutory but with focus on public transport or non-statutory climate change funded mobility strategies.
- Netherlands—the former system of advisory municipal traffic and transport plans (*GVVP*) is now obsolete, so instead non-statutory mobility vision documents were used.
- Sweden—non-statutory Local Transport Strategies (*Trafikstrategier*) were reviewed, and the transport policy sections of comprehensive land use plans (*Översiktsplaner*).

Some additional data about the transport and demographic characteristics of the cities are shown in Table 2. They are remarkably similar, the main differences being the much higher car use in British and two Swedish cities, higher car ownership in the smaller and particularly German cities, higher levels of cycling (replacing walking) in the Dutch cities and the higher rates of unemployment in the Swedish cities. (Note that mode share is for trips to work in British cities, but for all trips in others) These similarities increase the validity of the analysis, since they imply that actual problems are similar, so it is the way these same problems are viewed that instead influences the transport policies developed.

**Table 2.** Selected demographic and transport characteristics of cities.

| City | Population of Municipality (Year of Data) | Mode Share % All Trips by: | | | | Daytime PT Frequencies Per Hour Main Routes | Car Ownership Per 1000 People (2019) | % Population with Higher Education (Year) | % Workforce Out of Work 2019 |
|---|---|---|---|---|---|---|---|---|---|
| | | Car | PT | Bike | Walk | | | | |
| Aachen DE | 248,000 (2019) | 47 | 13 | 11 | 29 | 4 | 491 | 26 | 5.2 |
| Bath UK | 192,000 (2019) | 50 | 10 | 10 | 30 | 5 | 610 *(2011)* | 34 | 3.3 |
| Darlington UK | 100,500 (2019) | 72 | 9 | 3 | 16 | 4 | 510 *(2011)* | 20 | 5.6 |
| Eindhoven NL | 232,000 (2020) | 40 | 5 | 40 | 5 | 8 | 470 | 38 | 3.4 |
| Eskilstuna SE | 105,000 (2020) | 58 | 8 | 13 | 21 | 5 | 461 | 29 | 7.7 |
| Groningen NL | 231,000 (2020) | 44 | 5 | 33 | 18 | 6 | 350 | 40 | 5.4 |
| Herrenberg DE | 32,909 (2019) | 57 | 13 | 11 | 19 | 2 | 680 | 35 | 2.9 |
| Jönköping SE | 137,000 (2019) | 68 | 10 | 10 | 12 | 6 | 486 | 36 | 5.2 |
| Lindau DE | 25,253 (2019) | 49 | 6 | 27 | 18 | 2 | 750 | 32 | 2.3 |
| Lund SE | 120,000 (2020) | 42 | 16 | 26 | 16 | 5 | 522 | 55 | 9.0 |
| Malmö SE | 316,000 (2020) | 42 | 21 | 22 | 15 | 7 | 352 | 41 | 9.0 |
| Nottingham UK | 331,069 (2019) | 57 | 20 | 4 | 19 | 6 | 330 *(2011)* | 19 | 3.0 |
| Tilburg NL | 220,000 (2020) | 57 | 6 | 23 | 14 | 6 | 650 | 31 | 4.0 |

Note: For data sources see Appendix A. PT = public transport

## 4. Limitations

The approach chosen, based on documents only, limits the analysis to the field of policy content. Whilst policy documents will at times provide some details of the public consultation processes used to build their legitimacy, the full process of how the policy was produced is almost never described. This focus on content is a potential shortcoming of our work, but nonetheless we argue that a document analysis, particularly one theoretically anchored and comparing documents across countries, yields important results, the "why" of which could then be explored in further research.

Another potential shortcoming of our work it that the results are not statistically generalisable to a larger population of cities. As with the issue about the full process of how the policy was produced, we argue that it is the theoretically anchored arguments we make that will enable other researchers and practitioners to deduce whether or not the findings are analytically generalisable to other cities and countries. That our results are valid and relevant for other cities is shown by how [24,25] previous studies, which have used the same theoretical framework in analyses of a larger population of cities in one and the same country, have reached similar conclusions as we have done in this article.

It is a further limitation of this paper that these are all rather small cities in similar countries in northwest Europe. It could be argued that more meaningful results could be obtained from a comparison of very different cities, for example, from selecting cities in northern and southern Europe, or Europe and Asia. However, this limitation was a deliberate choice that the authors argue to be analytically important. The four countries from which cities were chosen have a lengthy history of structured transport planning in relation to objectives (see for example [69] for a review) that contrasts with a traffic engineering focused approach in most of the rest of the world. In addition, the similarity of their transport planning cultures means that the relatively subtle differences in problematisation that this paper seeks to analyse can be unpicked, whereas if cities with very different transport planning cultures were selected, a comparison would be much less valuable. Finally, these cities all produce some form of transport strategy document, which is not the case in many cities in the world. There is an extensive literature on spatial (if not transport) planning cultures (see for example [70]) and the four countries, and particularly Germany, Netherlands and Sweden, are identified within this literature as having similar planning cultures. Other researchers reviewing transport strategy documents have also limited their sample—in the case of [33], for example, to English language documents in EU capital cities. For all these reasons, the authors argue that it is wholly justified to deliberately limit the scope of this paper in this way.

Finally, whilst the documents reviewed are broadly similar in their purpose and the broad topics that they cover, they are absolutely not identical in structure nor in level of detail. Therefore, a wholly systematic analysis of identically structured documents that would be possible if, for example, reviewing the Local Transport Plans that were produced in England in the early 2000s according to quite rigid central government guidelines, was simply not possible in the case of the documents reviewed across cities and countries for this paper.

## 5. Cross-Case Comparison and Discussion

### 5.1. What Are the Problems that Are Highlighted?

The problems most typically highlighted in the strategies reviewed are *poor accessibility* (as a "bad" in itself, but also because it is seen to compromise economic growth); the negative impacts of traffic on liveability of the central part of the cities and therefore its ability to attract inhabitants, especially those needed to support a knowledge economy; local air and noise pollution; and road safety. This is exemplified by the way in which Swedish cities of Lund and Malmö describe how excesses in car use and space allotted to cars have produced unattractive city centres and unsustainable transport systems. This, they argue, has led to a lack of "urban cohesion"—a functionally mixed city that is an aesthetically attractive place to live, visit and shop—which, they argue, risks leading to lower economic growth and fewer jobs. In these two Swedish cities, the desire is to change conventional planning's focus on the car by no longer planning for (car) mobility but instead planning for accessibility. Similar ideas are found in Eindhoven, which seeks to create:

> An attractive living environment: the centre, residential and leisure areas are free of excessive traffic and there is high quality green and public space. It is pleasant to live, work and spend time ... [whilst] ... multimodal accessibility of the prime economic locations contribute to an excellent business climate. [42] (p. 7)

Bath is another city that can be used to illustrate the most common problematisation found in our sample cities:

> Bath will enhance its unique status by adopting measures that promote sustainable transport and reduce the intrusion of vehicles, particularly in the historic core. This will enable more economic activity and growth, while enhancing its special character and environment and improving the quality of life for local people. [36] (p. 5)

This concept of (poor) "accessibility" is one that is mentioned in most of the documents reviewed, but is not well-defined—in a sense, this is a shortcoming of the problematisations in many of the cities' transport strategies. In the scientific literature [71] (p. 1), it is defined as "the potential for reaching spatially distributed opportunities (for employment, recreation, social interaction, etc.)". However, the same authors acknowledge that whilst the general concept is widely used, there has been a historic disconnect in translating the scientific concept of accessibility into practice. It is not, therefore, surprising that in the documents reviewed, accessibility is at the same time both very important and poorly defined. As [72] (p. 236) point out in their study of how accessibility is treated in 32 metropolitan transport plans from North America, Europe, Australia and Asia, "there is a trend toward a greater integration of accessibility objectives in transport plans [but] ... plans need to have clearly defined accessibility goals" [emphasis added]. There is a further assumption in the documents reviewed, made without reference to scientific evidence, that if "accessibility" is threatened, this will have negative economic impacts. Furthermore, many but not all of the case study cities see congestion from cars as a threat to accessibility in certain parts of their geographical area, but mobility by car as a core element of guaranteeing accessibility in other parts. Overall, then, accessibility is a priority for cities but its causes and effects are poorly defined and sometimes appear to be in conflict.

While the threat to accessibility and therefore economic growth is the problem most typically highlighted in cities' strategies, there are small and economically very successful cities, such as Herrenberg and Lindau in Germany, that place less emphasis on transport contributing to economic growth than do the other 11 cities reviewed. Smaller cities also place more emphasis on the severance effect of major roads than do the larger cities—slightly paradoxically perhaps, but indicative of how major roads can be very significant in scale in terms of their contribution to the transport system of small towns.

In the "second rank" of problems in terms of how often they are mentioned come greenhouse gas emissions, public health and the occupation of public space by cars. These are mentioned by most cities but do not receive the same level of attention as the problems in the "first rank". In only a minority of cities is car use in itself seen as a problem; it is rather the consequences of "excessive" or "unnecessary" car use that are viewed as problematic. Table 3 summarises the main problems identified in the cities reviewed.

**Table 3.** Problems identified in each city.

| City | Accessibility Economic Impact | Noise and/or Local Air Pollution | Greenhouse Gases | Liveability | Road Safety | Social Inclusion | Health | Space for Cars | Car Use |
|---|---|---|---|---|---|---|---|---|---|
| Aachen DE | X | X | X | X | X | | X | X | X |
| Bath UK | X | X | X | X | X | X | X | X | X |
| Darlington UK | X | X | X | | X | X | X | | |
| Eindhoven NL | X | X | X | X | X | X | | X | |
| Eskilstuna SE | (X) | | X | | | | | | |
| Groningen NL | X | X | | X | X | | X | X | |
| Herrenberg DE | | X | X | | X | | | X | |
| Jönköping SE | X | X | X | | X | | | X | X |
| Lindau DE | X | X | X | X | X | X | | X | |
| Lund SE | X | X | X | X | X | | | X | |
| Malmö SE | X | X | X | X | X | X | X | X | |
| Nottingham UK | X | X | X | | X | X | X | | |
| Tilburg NL | Lack of a comprehensive view of problems | | | | | | | | |

Differences and Cities Lacking Explicit Policy Problematisations

Despite the similarities, there are differences between cities and their particular focus. For example, Groningen is quite unique in identifying parking on street as a problem of *privatisation* of public space, although many cities see the use of street space for parking as an opportunity cost. All the British cities and the Swedish city Malmö stress social exclusion as a transport-related problem; in other cities and countries, there is much less focus on this issue, although several others do mention their ageing populations and the need to ensure that the transport system caters to the needs of older people. This can be explained by how economic downturn and population decline in the British cities and in Malmö the last decades have led to a focus on social exclusion as a problem and on an improved transport system as a remedy.

Four cities (Eindhoven, Eskilstuna, Herrenberg and Tilburg) do not, at least in the plans we have examined, provide explicit problematisations; rather, the transport-related problems they suffer are only implied via the choice of objectives in the strategies (all strategies include a statement of objectives). There appears to be no link between not identifying problems explicitly and failing to have reduction in car use as an objective in Tilburg and Herrenberg. Tilburg is a city where there is no problematisation and there is also no objective to reduce car use; in contrast, Herrenberg's strategy, in spite of having no problematisation, nonetheless is very explicit that car use should be reduced.

To conclude, differences between cities clearly exist, and they seem partly to be related to city differences in terms of economic development. Nevertheless, there are significant similarities between cities regardless of country and similar policy problematisations. Importantly, policy problematisations often have a geographical delimitation, with a focus on cars in inner city and city centres as a problem that compromises the "accessibility" of these areas. We describe this in detail in the next section when linking policy problematisations to mobility futures imagined. There is less of an explicit process of policy problematisation regarding outer city and regional transport. It must be assumed that this is a result of how the problem of cars is defined, i.e., that the major problem of cars arises in inner city and city centres.

### 5.2. Mobility Futures Imagined—Key Elements, Commonalities and Differences

The cities that explicitly state that, in their respective future mobility systems, car use will be lower than it is today are Lindau, Lund, Aachen, Bath, Nottingham, Herrenberg and Eindhoven—so the majority of the 13 studied. Lindau, Lund, Bath and Herrenberg fall into the category of "smaller historic city", but on the other hand, the three other cities are mixed industrial/university cities and all amongst the largest cities studied for this research.

Other than this, there are considerable commonalities in the mobility futures imagined, particularly with respect to the central city, where, in general in pursuit of greater "liveability" cities, imagine streets where more space is given to sustainable modes of transport and used as public space, and where car use, even if not reduced, is directed to locations such as underground car parks where its immediate impacts are lessened compared to the current situation, as shown in Table 4. All cities imagine futures where at least in central areas more of the available on-street parking is reduced and, even if there is no explicit recognition of a need to reduce car use, nonetheless the implication is that, in these areas, travel by car will become relatively less important as inner-city populations grow. Even though the extension of car traffic in the cities are often explicitly seen as a problem in the strategies, and constructed as a condition that needs to be changed by means of interventions in future planning, it is not necessarily a completely different transport future that is imagined, below exemplified by Jönköping (SE) and Groningen (NL):

**Table 4.** Elements of mobility futures imagined identified in each city.

| City | Car reduced City Centre, with Further Reallocation of Space from Car | Reduced on-Street Parking | Cycle Routes | New Road Building Supported/ Planned | Cut Road Space for Car Outside Centre | Maintain Car Accessibility | Better Public Transport | Lower Speed Limits | Explicit Aim to Reduce Levels of Car Use |
|---|---|---|---|---|---|---|---|---|---|
| Aachen DE | X | X | X | | X | | X | X | X |
| Bath UK | X | X | X | | X | | X | X | X |
| Darlington UK | | | X | X | | X | X | X | |
| Eindhoven NL | X | X | X | | X | | X | X | X |
| Eskilstuna SE | | | X | X | | X | X | | |
| Groningen NL | X | X | X | X | | X | X | | |
| Herrenberg DE | X | X | X | X | | X | X | X | X |
| Jönköping SE | | X | X | X | | X | | | X |
| Lindau DE | | | X | X | | X | X | X | X |
| Lund SE | X | X | X | | | | X | X | X |
| Malmö SE | X | X | X | | | | X | X | |
| Nottingham UK | X | X | X | X | X | | X | X | X |
| Tilburg NL | X | X | X | X | | X | X | X | |

> The majority of all trips made in the municipality are made by car, both now and in the future. … Roughly 20% of all car trips could be replaced with other options. … Such a change in travel mileage would entail that those car trips that cannot be replaced with other, more sustainable options could also continue to offer free flowing travel conditions. [53] (p. 52)

> In all future scenarios the car remains as a dominant mode and a solid basic vehicle infrastructure is therefore necessary. [48] (p. 32)

Most cities thus believe that the car will continue to play an important role in the transport system. In other words, a change including radical reductions in car use is often not imagined. Instead, as populations grow and pressure on transport systems increases, there is a desire to make it possible for those who wish to drive their cars to the central parts of cities with ease to continue to do so in the future. The ambition is more about making the transport system more efficient by cutting the space allotted to cars within it, rather than radically reducing car use. Access by car is to be restricted in only three or four cities, including Eindhoven, Aachen and (in the shape of demand management) Nottingham.

### 5.3. *What Measures Are Typical and Less Typical in These Mobility Futures?*

The Dutch cities, in particular, are very enthusiastic about the use of big data, self-driving vehicles, new forms of mobility (such as car sharing [73]) and electric mobility, although it is not always clear how big data and self-driving vehicles will be used as measures to further the objectives of the strategies. The majority of strategies reviewed include measures such as charging points and some electrification of public transport and municipal vehicle fleets in order to help meet pollution reduction objectives, but these are not a major feature of the majority of the strategies reviewed. The Dutch municipalities are all proponents of inter-city cycle superhighways, something not mentioned in strategies from other countries. The importance of park and ride diminishes with city size, in general. However, other than these points, there is great commonality between all cities in the nature of the measures they propose: management of on-street parking, some conversion of street space to public space, improved conditions for walking, improved public transport and much improved cycle networks and cycle parking (see Table 3).

These measures that apply mainly to other modes of transport *could* have a direct impact on car traffic, for example, if walking, cycling and public transport are made more attractive relative to car use, by redistributing street space in favour of pedestrians, bicyclists and public transport [15]. However, this reallocation of space is discussed explicitly in only a minority of cities. While certain cities, such as Tilburg, Groningen, Lindau, Herrenberg, Darlington and Eskilstuna, are explicit about road improvements that will take place during the lifetime of their strategy on the local and/or national road network in their areas, only a few cities, such as Eindhoven and Aachen, rule out new roads, or additional road capacity, completely. Eindhoven, Aachen, Nottingham and Bath are more explicit about managing demand through reallocating road space (and through the use of the workplace parking levy in the case of Nottingham) than are any of the other cities examined. For example, Eindhoven's strategy talks positively about making certain key arterials one-way streets in order to free up space for other uses; Bath's strategy strongly supports further pedestrianisation in the city. These points are summarised in Table 3, which shows that the key similarities and differences between the mobility futures imagined are:

- Almost all cities plan to reallocate road space away from cars in their central areas and to manage on-street parking.
- At the same time, most want to maintain car access to these areas (meaning that car access will not be restricted but parking will be more expensive and/or off-street).
- All cities plan to improve conditions for cyclists and most conditions for public transport.

- A small number of cities plan to reallocate road space away from cars outside their central areas.
- The majority of cities plan or support road capacity increases within their areas.

Finally, the majority of land use plans examined are also supportive of transport policy goals, and vice versa—an integration that [2,74] previous research has shown to be of strategic importance when decreasing the need to travel by car and make it rational to choose public transport, walking or cycling. For example, a reduction of car travel in Lund [58] (p. 13), [59] (p. 37) is planned to result from the integration of land use and transport planning (described as urban planning) that results in densification and a mix of functions in locations with good conditions for public transport. Similarly, Aachen's land use plan is clear about the need to increase densities and to locate new residential and commercial developments along corridors and at nodes with good public transport service, rather than on promoting low-density, car-based development. It also envisages some controls on out-of-town shopping in order to maintain and strengthen the role of the city centre as a retail destination [35,36].

Due to the timeframe of the documents reviewed, the most recent being from 2017, the issue of COVID-19 and its implications for mobility futures was not considered in any of the policy documents reviewed. If repeated five years hence, our review might find policy documents that pay more attention to providing space for social distancing, to increased reliance on personal micromobility instead of public transport, to increased use of virtual mobility and to greater uncertainty in planning inputs and outcomes. However, this is obviously highly speculative at this point in time.

### 5.4. *What Ambiguities/Conflicts/Silences Are There in the Imagined Mobility Futures?*

The most significant ambiguity in the majority of the cases studied is whether or not the scale of the measures proposed, particularly the demand management and road space reallocation measures, is sufficient to meet the level of change in travel behaviour to which most cities aspire. In addition, where cities support the addition of new road capacity on their own roads or on those of national authorities, then there is some ambiguity in that this new capacity may induce more car travel that will then undermine the achievement of other objectives in the strategies.

A related point is the view that most strategies take on regional, national and international travel, for which in the majority of plans there is no desire, either explicit or implicit, to reduce car use; it is seen as a necessity, yet the ease with which this could undermine the achievement of sustainable mobility objectives at the local level is not acknowledged. The principal conflict is between a more car-based mobility future in suburbs and regions around cities, and restraint of car use for a higher quality of life in central cities, with no recognition that the former could well undermine the latter. Regional car use is often not seen by cities as a "problem" to be solved, and some, such as Groningen, even explicitly recognise the need to maintain and even increase car use in these areas. However, there are exceptions, such as Aachen, which envisages a mobility future where, in their entire area, sustainable modes of transport become the most important modes.

The unproblematised role of regional car use in transport policy can be understood in the light of the assumption some cities make about how economic development depends on a high degree of intraregional mobility among the cities in their region. An underlying assumption appears to be that spatially integrated cities, through increased regional *accessibility* (the potential to reach jobs and services regionally) will lead to opportunities for both employers and employees to choose labour and work. The mobility future imagined on a regional scale is thus a future with "high mobility-high accessibility". For example, the Mobility Vision of the Dutch city of Tilburg aspires to improvements in all modes of transport at the regional scale, whilst the City of Groningen states that [51] (p. 22) "further intensification of land uses in the A7 motorway corridor is only possible if the traffic infrastructure around it grows at the same time". Whilst the historic city of Bath rules out new road building within the city, there are many proposals for new road as well as public transport capacity increases at the level of the region within which Bath sits, again evidence of a high mobility-high accessibility strategy in this English case.

Finally, there are major silences related to freight transport and deliveries; few strategies (only from Aachen, Lund, Malmö and Tilburg) explicitly mentioned this issue.

The cities reviewed for this paper that have a reputation for being more "ambitious" in their transport policies are Lund, Malmo, Nottingham, Groningen and possibly Lindau. From the documents reviewed, it is not possible to say that these cities are all "stronger" in terms of either their mobility futures imagined or the measures that they propose to bring about these futures, than other cities that feature as cases in this study; there is no consistent pattern. So, for example, Nottingham is the only city that employs a measure other than parking management and road space reallocation to restrain car traffic, but at the same time it proposes a major road infrastructure scheme. Aachen in Germany, although not a city with a reputation for radical transport policies, nonetheless describes a radical mobility future and measures to achieve it.

From our problematisation-based analysis, it is *not* possible to conclude that those cities with the greatest ambiguities and silences in their policy packages (for example, Tilburg, Groningen, Lindau or Darlington) are those that also have significantly different problematisations from those such as Aachen or Eindhoven that have few ambiguities and silences. The conception of problems is similar, but differences arise at the level of responses to those problems; however, a document analysis does not reveal the reasons for those differences.

To conclude, the key factor excluded in the most mobility futures is how mobility is to be provided for in the more suburban areas of the cities studied. Similar silences exist for inter-regional trips. There are also differences regarding the degree to which any demand management measure other than parking management (such as road space reallocation, or some form of pricing) is included in the measures in the strategy. Statements about encouraging sustainable modes of transport are common in the mobility futures, but it is only in a minority that there is an acceptance of reducing road capacity for private vehicles. By far the most common demand management tool included in the strategies reviewed is management of on-street parking, but many cities at the same time make no commitment to reduce off-street parking. Some mobility futures imagine unfettered car access even to central cities, for those car users willing to pay higher rates for parking.

Table 5, below, summarises the findings of Section 5.4.

**Table 5.** Analysis of problem representations, comparison cities.

| City | Problem with Cars/? | Mobility Future Imagined? | Assumptions Underlying This Representation of the Problem: Key Concepts; Actions Required | Unproblematic Issues/Silences in This Representation of the Problem | Potential Consequences Produced |
|---|---|---|---|---|---|
| Aachen DE | Car use is threat to economic development; pollution; quality of life; road safety. | Much less car dependent and car dominated including in wider region. | Assumptions: relationship between accessibility, quality of life and economic development; Measures: reallocation of road space to sustainable modes plus parking management. | Will traffic restraint measures will be enough to reduce car use? | If restraint measures are not sufficient, car use will to grow, undermining measures in central city. |
| Bath UK | Car use as threat to quality of life; protection of historic environment; road safety; economic development; wider environment. | Less car dependent within central city at least but not wider region. | Assumptions: relationship between accessibility, quality of life and economic development; Measures: reallocation of road space to sustainable modes plus parking management. Land use to be planned to support sustainable transport. | Will traffic restraint measures will be enough to reduce car use? Interrelationship between city level and regional level Managing freight transport. | As above.<br>Also, conflict between city level and regional level. |
| Darlington UK | Poor accessibility as threat to economic development, to social inclusion, to safety and to environment. | Continuation of existing situation but with increased choices and no constraint on car use. | New road infrastructure very important in mobility future. Little attempt to link land use and sustainable transport. Assumption that transport system can continue broadly "as is". | How lack of measures to restrain car use is consistent with problems identified. | Problems caused by car use unlikely to be tackled. |
| Eindhoven NL | Poor accessibility as threat to quality of life and thus to economic development, to safety and to environment. | Increasing choice of modes and better accessibility, but car takes much-reduced role within this at all geographical scales, not only in central city. | Assumptions: relationship between accessibility, quality of life and economic development; Measures: reallocation of road space to sustainable modes plus parking management. Land use to be planned to support sustainable transport. | Will traffic restraint measures will be enough to reduce car use? Interrelationship between city level and regional level Managing freight transport. | If restraint measures are not sufficient, car use will continue to grow, undermining measures in central city especially. |

**Table 5.** *Cont.*

| City | Problem with Cars/? | Mobility Future Imagined? | Assumptions Underlying This Representation of the Problem: Key Concepts; Actions Required | Unproblematic Issues/Silences in This Representation of the Problem | Potential Consequences Produced |
|---|---|---|---|---|---|
| Eskilstuna SE | Vaguely expressed. Main problem: continuation of economic and demographic decline. | Mobility futures imagined–"High mobility-high accessibility". | Key assumptions relate to relationship between accessibility, quality of life and economic development. | Interrelationship between city level and regional level. | Increasing number of trips due to the vision of increased accessibility and mobility. |
| Groningen NL | Poor accessibility as threat to quality of life and thus to economic development, to safety and to environment. Opportunity cost of space used by cars. | Increasing choice of modes and better accessibility and car takes much-reduced role within this but only in central city. | Assumptions: relationship between accessibility, quality of life and economic development; Measures: reallocation of road space to sustainable modes plus parking management in central city. Outside that, key measures include improved public transport, long distance cycle routes and increases in road infrastructure. | Inconsistency between restraint measures in central area and measures for increased transport supply in outer city and wider region.<br>Freight transport. | Risk that strategy for outer city will undermine that for central city. |
| Herrenberg DE | Safety; Local and global pollution; Road safety; Severance caused by major roads. | High accessibility by sustainable modes<br>Improved quality of life in city centre especially. | Car use should be reduced<br>Policy fields such as pollution reduction, land use and mobility are all mutually interlinked<br>There is a link between transport and the quality of life in the city centre. | How to manage regional car trips—Will traffic restraint measures will be enough to reduce car use?<br>Links to the surrounding areas. | Risk that strategy for outer city and regional trips will undermine that for central city. |
| Jönköping SE | Volume of car traffic restricts public transport, leading to traffic safety and air quality problems, noise and climate change. | Car traffic has to give room to sustainable modes of travel.<br>However, a radical change is not imagined.<br>The majority of travels will be by car now as well as in the future. | The importance of individuals and choice of transport modes are emphasised, leading to a focus influencing travel behavior through "soft governance", such as information and offering alternatives to the car. | Regional transport, "hard measures", like changing the physical space for cars in the city. | Car use will continue to grow. Problems caused by car use unlikely to be tackled. |

**Table 5.** *Cont.*

| City | Problem with Cars/? | Mobility Future Imagined? | Assumptions Underlying This Representation of the Problem: Key Concepts; Actions Required | Unproblematic Issues/Silences in This Representation of the Problem | Potential Consequences Produced |
|---|---|---|---|---|---|
| Lindau DE | Pollution, safety, congestion, severance, impact of vehicles on historic environment. | High % of trips by sustainable modes, all areas, but within a highly mobile region. Car traffic still very important for at least trips transiting the city. | More control of parking for private cars required across city. Improvement of alternative modes. Road safety black spot approach. Capacity enhancements at major road junctions. | How to manage regional car trips—Will traffic restraint measures will be enough to reduce car use? Contradiction between demand management and road capacity improvements. Freight traffic and deliveries. | Aspirations for a more liveable city where sustainable transport carries majority of trips undermined/threatened by unproblematic issues, especially capacity enhancements. |
| Lund SE | Emissions (particularly greenhouse gas emissions); Negative effects on living environments. | Sustainable transport system; Attractive living environments. | Planning for *accessibility* (not for mobility); *Priority* for walking, cycling and public transport; Increasing *attractiveness* of environmentally friendly transport modes; *Integrated* land use and transport planning. | None, compared to other cities. | Some problems caused by car use (negative effects on living environments) likely to be tackled. |
| Malmö SE | Justice, equity, inclusion, accessibility; climate. | Sustainable urban mobility with emphasis on social sustainability. Walking, cycling, public transport is the "natural" choice. | Key measures include improved conditions for walking, cycling and public transport; street will have to be transformed, space reallocated. | None, compared to other cities. | Some problems caused by car use likely to be tackled. |
| Nottingham UK | Poor accessibility as threat access to jobs and thus to economic development, to safety and to environment. Social inclusion and road safety. | Future imagined is one where alternative modes predominate for all trips especially in more central city. | Assumptions: relationship between accessibility and economic development; Measures: reallocation of road space to sustainable modes plus parking management, improved public transport and one major urban road building scheme. | Will traffic restraint measures will be enough to reduce car use? Whether improvements in transport will improve social inclusion. | Possibility that planned major road investment (ring road) will undermine achievement of other objectives. |

**Table 5.** *Cont.*

| City | Problem with Cars/? | Mobility Future Imagined? | Assumptions Underlying This Representation of the Problem: Key Concepts; Actions Required | Unproblematic Issues/Silences in This Representation of the Problem | Potential Consequences Produced |
|---|---|---|---|---|---|
| Tilburg NL | Poor accessibility as threat to economy; Road safety; Local and global pollution; Deteriorating quality of life due to increased car use. | High accessibility by all modes; Improved quality of life in city centre especially. | Car use is still a given and cannot or should not be reduced<br>Improved accessibility by all modes is achievable and one mode does not conflict with another<br>Regional connectivity must be provided by road and is necessary for economic growth. | How to manage regional car trips—Will traffic restraint measures will be enough to reduce car use?<br>Whether improving traffic flow for private vehicles will induce more vehicle trips onto network. | Risk that strategy for outer city and regional trips will undermine that for central city. |

## 6. Conclusions

An analytical framework based on [6,7] Bacchi's work on policy problematisation was used for this work. Some of the limitations of the framework were discussed earlier in this paper—principally that that the focus on problems as the foundation of policy-making ignores the possibility that solutions and measures are already selected and thus the definition of the problem is used as a post-hoc justification of the already-chosen measures. However, since this paper is based on a document analysis, it would be difficult to identify a post-hoc justification of the measures selected by cities, and therefore a problematisation-based analytical framework lends itself better to such an analysis. In addition, we present below a number of important conclusions regarding the documents that we have analysed and, as these stem from our analytical framework, that demonstrates its usefulness.

There is of course other research that has analysed the urban mobility strategies of cities (see for example [74–78]). However, only one of these papers undertakes an international comparison of more than four strategies, and all are focused on the quantitative indicators used in the plans to measure their success. None analyses the problem statements in the plans in depth as this paper has done, and none identifies the conflicts within the mobility futures imagined. Therefore, our paper complements this earlier work, but clearly also adds to it.

It is clear from our analysis that the majority of the cities reviewed have similar conceptions of the problems caused by or are related to car use. The relative importance of these problems varies somewhat according to the social and economic context of the city, but in general the key problems are *poor accessibility as a threat to economic growth*, followed by local then global environmental impacts and road safety problems. The majority of the cities also state that they seek to reduce car use as a proportion of trips, with a minority of these setting a quantitative target for this. More broadly, the desired mobility future seen in most cities is one where the use and dominance of the private car are reduced, leading to improved public space and liveability, particularly in the central part of cities. The mobility future in all cities envisages a broader range of mobility options compared to the situation today.

The key differences between cities lie in the degree to which the measures that will restrain and reduce private car use are emphasised. Some cities are more ready to use such measures than others, and some make it clear in addition that car use will remain the dominant mode of transport in their mobility system. Furthermore, there are differences in the degree to which car restraint measures are planned to be used in the outer city and regionally, as compared to the central city. Whilst there are exceptions, *the general pattern is that restraint measures are mostly envisaged for the central city, whilst it is planned that the outer city and region will benefit from increased mobility by all modes.*

This clearly leads to a number of "silences" and ambiguities between the policy problematisation and the measures proposed. Principally, these include the lack of clarity as to whether car-restraint measures will be sufficient to actually reduce car use, and *the conflict between measures proposed for the central city and those proposed for areas further out from the central city.* These silences are of great importance for the transition towards sustainable transport systems in terms of $CO_2$ emissions and overall km travelled.

The paper did not set out to explain *why* the patterns identified are as they are. However, the issue of path dependence referred to in the introduction is clearly relevant (e.g., [1,8,9,21] regarding path dependence in transport planning). There is a continuum of cities here in the 13 considered. Some remain in an "old" traffic engineering-based paradigm, with only a few cursory "nods" towards sustainable transport in their policies. The majority have moved some way towards a more thoroughly sustainable transport paradigm, where cars are not part of their vision for the central city. A very small number are rejecting the car as a basis for future mobility for the whole city. This is a path-dependent transition, as cities in the first stage are unlikely to jump straight to the last, without going through the intervening stage).

In addition, our work highlights at least two important potential consequences for urban transport policy in general:

- Firstly, the policies that we have reviewed suggest that if they are implemented as planned, suburbs may in fact become more car-dominated. The resulting increase in vehicle kilometres due to regional car trips is likely to increase local and global air pollution until a high proportion of the fleet is electrified. This is a silence that could have major consequences for the energy efficiency of the transport system since these trips account for a large part of total personal transportation mileage.
- Secondly, car mobility is rarely put at stake in the design of policy interventions. Few measures that directly aim to restrict car use are presented. Possible measures that will drastically reduce car use, such as car-free urban areas, are ruled out.

For planners working in cities, the analysis in this paper is an important call for them to recognise the silences in their own policy documents, and for them to think more critically about how to manage suburban and regional car use. For researchers, our work should encourage a more critical view to be taken of the discourses in urban mobility strategies, particularly with respect to how conflicts in the problematisations in those strategies are dealt with.

In summary, it is true that the cities analysed in this paper have taken some potentially important steps towards sustainable transport planning (principally in the planning of central parts of the cities, that is). However, the dominant policy problematisation, with its geographical delimitation focusing on cars in city centres, produces silences that risk segmenting car-based mobility futures in suburbs and at the regional level. In order to at all be able to handle these car journeys, cities need explicit policy problematisations for car journeys in suburbs and at the regional level. If not, these car journeys will not be dealt with as they are discursive blind spot.

**Author Contributions:** T.R.: introduction, methodology, first analysis of plans outside Sweden. R.H.: literature review, first analysis of Swedish plans. T.R. and R.H.: discussions and conclusions. All authors have read and agreed to the published version of the manuscript.

**Funding:** This research was carried out in collaboration with K2—The national Swedish Knowledge Centre for Public Transport www.k2centrum.se. It was funded by K2 and the Swedish Energy Agency (grant number 43202-1), which had no involvement in the study design, in the data collection, analysis, and interpretation, or in the writing of the paper.

**Conflicts of Interest:** The authors declare no conflict of interest.

## Appendix A Sources of Data, Table 2

City populations: city authority websites, latest available year

Mode share for Dutch and Swedish cities: latest available year from http://www.epomm.eu/tems/cities.phtml

Mode share for German cities: from transport strategies reviewed, for 2018 (2017 Aachen).

Mode share for British cities: 2011 National Census travel to work data.

Education levels, car ownership and unemployment rates:

Britain: 2011 census (education and car ownership); city websites (unemployment).

Germany: car ownership from transport strategies reviewed. Education and unemployment from Employment Ministry data of respective German Federal State

Netherlands: Central Bureau for Statistics (CBS) (Dutch Government)

Sweden: CBS (Sweden) for unemployment and education; Lansstyrelsen for car ownership.

## References

1. Low, N.; Gleeson, B.; Rush, E. Making Believe. Institutional and Discursive Barriers to Sustainable Transport in Two Australian Cities. *Int. Plan. Stud.* **2003**, *8*, 93–114. [CrossRef]
2. Næss, P.; Hansson, L.; Richardson, T.; Tennøy, A. Knowledge-based land use and transport planning? Consistency and gap between 'state-of the-art' knowledge and knowledge claims in planning documents in three Scandinavian city regions. *Plan. Theory Pract.* **2013**, *14*, 470–491. [CrossRef]

3. Isaksson, K.; Antonson, H.; Eriksson, L. Layering and parallel policy making—Complementary concepts for understanding implementation challenges related to sustainable mobility. *Transp. Policy* **2017**, *53*, 50–57. [CrossRef]
4. Hrelja, R. The tyranny of small decisions. Unsustainable cities and local day-to-day transport planning. *Plan. Theory Pract.* **2011**, *12*, 511–524. [CrossRef]
5. Richardson, T.; Isaksson, K.; Gullberg, A. Changing frames of mobility through radical policy interventions? The stockholm congestion tax. *Int. Plan. Stud.* **2010**, *15*, 53–67. [CrossRef]
6. Bacchi, C. *Women, Policy and politics: The Construction of Policy Problems*; Sage: London, UK, 1999.
7. Bacchi, C. *Analysing Policy. What's the Problem Represented to Be*; Pearson Australia: Frenchs Forest, Australia, 2009.
8. Hensley, M.; Mateo-Babiano, D.; Minnery, J. Healthy places, active transport and path dependence: A review of the literature. *Heal. Promot. J. Aust.* **2014**, *25*, 196–201. [CrossRef] [PubMed]
9. Low, N.; Astle, R. Path dependence in urban transport: An institutional analysis of urban passenger transport in Melbourne, Australia, 1956–2006. *Transp. Policy* **2009**, *16*, 47–58. [CrossRef]
10. Santos, G.; Behrendt, H.; Teytelboym, A. Part II: Policy instruments for sustainable road transport. *Res. Transp. Econ.* **2010**, *28*, 46–91. [CrossRef]
11. Marshall, S.; Banister, D. Travel reduction strategies: Intentions and outcomes. *Transp. Res. Part A Policy Pract.* **2000**, *34*, 321–338. [CrossRef]
12. Marshall, S. Introduction. Travel reduction—Means and ends. *Built Environ.* **1999**, *25*, 89–93.
13. Buehler, R.; Pucher, J. International sustainable tansport in Freiburg. Lessons from Germany's environmental capital. *J. Sustain. Transp.* **2011**, *5*, 43–70. [CrossRef]
14. Buehler, R.; Pucher, J.; Altshuler, A. Vienna's path to sustainable transport. *Int. J. Sustain. Transp.* **2017**, *11*, 257–271. [CrossRef]
15. Buehler, R.; Pucher, J.; Gerike, R.; Götschi, T. Reducing car dependence in the heart of Europe: Lessons from Germany, Austria, and Switzerland. *Transp. Rev.* **2016**, *37*, 4–28. [CrossRef]
16. Hysing, E. Greening transport—Explaining urban transport policy change. *J. Environ. Policy Plan.* **2009**, *11*, 243–261. [CrossRef]
17. Nikulina, V.; Simon, D.; Ny, H.; Baumann, H. Context-adapted urban planning for rapid transitioning of personal mobility towards sustainability: A systematic literature review. *Sustainability* **2019**, *11*, 1007. [CrossRef]
18. McTigue, C.; Monios, J.; Rye, T. Identifying barriers to implementation of local transport policy: An analysis of bus policy in Great Britain. *Util. Policy* **2018**, *50*, 133–143. [CrossRef]
19. Marsden, G.; Reardon, L. Questions of governance: Rethinking the study of transportation policy. *Transp. Res. Part A Policy Pract.* **2017**, *101*, 238–251. [CrossRef]
20. Fischer, F.; Torgerson, D.; Durnová, A.; Orsini, M. *Handbook of Critical Policy Studies*; Edward Elgar Pub.: Northampton, MA, USA, 2016.
21. Low, N.; Gleeson, B.; Rush, E. A multivalent conception of path dependence: The case of transport planning in metropolitan Melbourne, Australia. *Environ. Sci.* **2005**, *2*, 391–408. [CrossRef]
22. Foucault, M. *The History of Sexuality*; Penguin Books: London, UK, 1976; Volume 1.
23. Rein, M.; Schon, D. Frame-critical policy analysis and frame-reflective policy practice. *Knowl. Soc.* **1996**, *9*, 85–104. [CrossRef]
24. Hrelja, R. Cars. Problematisations, measures and blind spots in local transport and land use policy. *Land Use Policy* **2019**, *87*, 104014. [CrossRef]
25. Rehnlund, M. *Getting the Transport Right—For What. What Transport Policy Can Tell Us about the Construction of Sustainability, Dissertation*; Södertörns högskola: Huddinge, Sweden, 2019.
26. Kingdon, J.W. *Agendas, Alternatives, and Public Policies*; Little, Brown: Boston, MA, USA, 1984.
27. Jones, M.D.; Peterson, H.L.; Pierce, J.J.; Herweg, N.; Bernal, A.; Lamberta Raney, H.; Zahariadis, N. A river runs through it: A multiple streams meta-review. *Policy Stud. J.* **2016**, *44*, 13–36. [CrossRef]
28. Campisi, T.; Akgün, N.; Ticali, D.; Tesoriere, G. Exploring public opinion on personal mobility vehicle use: A case study in Palermo, Italy. *Sustainability* **2020**, *12*, 5460. [CrossRef]
29. Salminen, A.; Kauppinen, K.; Lehtovaara, M. Towards a methodology for document analysis. *J. Am. Soc. Inf. Sci.* **1997**, *48*, 644–655. [CrossRef]
30. Bowen, G.A. Document analysis as a qualitative research method. *Qual. Res. J.* **2009**, *9*, 27. [CrossRef]

31. Hill, A.; Song, S.; Dong, A.; Agogino, A. Identifying shared understanding in design using document analysis. In Proceedings of the 13th International Conference on Design Theory and Methodology, September 2001; American Society of Mechanical Engineers, New York, NY, USA; pp. 9–12.
32. Flyvberg, B.; Flyvbjerg, B. Five misunderstandings about case-study research. *Qual. Inq.* **2006**, *12*, 219–245. [CrossRef]
33. Kiba-Janiak, M.; Witkowski, J. Sustainable urban mobility plans: How do they work? *Sustainability* **2019**, *11*, 4605. [CrossRef]
34. City of Aachen Mobility Vision 2050. 2014. Available online: http://www.aachen.de/de/stadt_buerger/verkehr_strasse/verkehrskonzepte/VEP/Vision2050/VisionMobilitaet_2050_neu.pdf (accessed on 4 February 2020).
35. City of Aachen Mobility Strategy 2030 (Verkehrsentwicklungsplan (VEP)). 2015. Available online: http://www.aachen.de/de/stadt_buerger/verkehr_strasse/_materialien_verkehr_strasse/verkehrskonzepte/vep/Strategie2030/15-06-11-Plakat-Innenstadt.pdf (accessed on 4 February 2020).
36. Spatial plan policy section (Flachennutzplan). 2014. Available online: http://www.aachen.de/DE/stadt_buerger/planen_bauen/stadtentwicklung/stadt/aachen2030/fundus/index.html (accessed on 4 February 2020).
37. Bath and Northeast Somerset Council. Getting Around Bath—A Transport Strategy for Bath. 2014. Available online: https://www.bathnes.gov.uk/sites/default/files/sitedocuments/getting_around_bath_transport_strategy_-_final_issue_web_version.pdf (accessed on 6 February 2020).
38. Bath and Northeast Somerset Council. Draft joint, regional, Local Transport Plan (LTP) number 4. 2019. Available online: https://travelwest.info/projects/joint-local-transport-plan (accessed on 6 February 2020).
39. Bath and Northeast Somerset Council. Core Strategy of the Local Plan. 2014. Available online: https://www.bathnes.gov.uk/sites/default/files/sitedocuments/Planning-and-Building-Control/Planning-Policy/Core-Strategy/InfoPapersandAppraisals/core_strategy_sa_report_2014.pdf (accessed on 6 February 2020).
40. Bath and Northeast Somerset Council. The Placemaking Strategy. 2017. Available online: https://beta.bathnes.gov.uk/policy-and-documents-library/core-strategy-and-placemaking-plan (accessed on 6 February 2020).
41. Darlington Borough Council. Third Local Transport Plan. 2011. Available online: https://www.darlington.gov.uk/media/2873/third-local-transport-plan.pdf (accessed on 5 February 2020).
42. Darlington Borough Council. Core Strategy of the Local Plan. 2018. Available online: https://www.darlington.gov.uk/media/11443/corestrategy-dpd.pdf (accessed on 5 February 2020).
43. City of Eindhoven. Mobility Vision 2040, (Eindhoven op Weg). 2013. Available online: https://www.eindhoven.nl/sites/default/files/2019-04/2014-04-15%20Eindhoven%20op%20weg%20REPRO.pdf (accessed on 5 February 2020).
44. City of Eindhoven. New Parking Norms (Nota Parkeernormen). 2016. Available online: https://www.planviewer.nl/imro/files/NL.IMRO.0772.80282-0301/b_NL.IMRO.0772.80282-0301_rb3.pdf (accessed on 5 February 2020).
45. City of Eindhoven. The Land Use Strategy for housing (Beleidsnota Prioriteiten bouwlocaties). 2015. Available online: https://www.eindhoven.nl/sites/default/files/2018-05/Beleidsnota%20Prioriteiten%20bouwlocaties%202015_VASTGESTELD.pdf (accessed on 5 February 2020).
46. Eskilstuna Municipality. Trafikplan för Eskilstuna Municipality 2012. ***Strategidel.*** 2012. Available online: https://www.eskilstuna.se/download/18.548e06b0158e46395f66087/1482151068950/Trafikplan%20-%20strategi.pdf (accessed on 10 February 2020).
47. Eskilstuna Municipality. Översiktsplan 2030. 2013. Available online: https://www.eskilstuna.se/bygga-bo-och-miljo/stadsplanering-och-byggande/stadsplanering/oversiktsplanering/oversiktsplan-2030.html (accessed on 10 February 2020).
48. City of Groningen. Mobility Strategy (Nota Mobiliteit) 2007–2010. 2007. Available online: http://decentrale.regelgeving.overheid.nl/cvdr/xhtmloutput/actueel/Groningen%20(Gr)/CVDR100158.html (accessed on 10 February 2020).
49. City of Groningen. Multi-year Programme for Transport and Traffic 2018–2021 (Meerjarenprogramma Verkeer 2018–2021). 2017. Available online: https://gemeenteraad.groningen.nl/Documenten/Raadsvoorstellen/Bijlage-Meerjarenprogramma-Verkeer-en-Vervoer-2018-2021-2.pdf (accessed on 10 February 2020).

50. City of Groningen. Parking Strategy (Ruimte voor de Straat). 2018. Available online: https://gemeente.groningen.nl/sites/default/files/Ruimte-voor-de-straat---parkeren-in-een-levende-stad.pdf (accessed on 10 February 2020).
51. City of Groningen. Spatial Planning Strategy (Nota Grondbeleid). 2017. Available online: https://gemeenteraad.groningen.nl/Documenten/Raadsvoorstellen/Nota-Grondbeleid-2017 (accessed on 10 February 2020).
52. City of Herrenberg. Integrated Mobility Development Plan (Integrierter Mobilitätsentwicklungsplan Stadt Herrenberg). 2019. Available online: https://www.herrenberg.de/tools/partPlat/projects/pdfs/37/B6xkOJ (accessed on 10 February 2020).
53. Jönköping Municipality. Kommunikationsstrategi. 2012. Available online: https://www.jonkoping.se/download/18.74fef9ab15548f0b80012aa0/1465889660350/Kommunikationsstrategi%20-%20%C3%A5tg%C3%A4rder%20f%C3%B6r%20ett%20h%C3%A5llbart%20trafiksystem.pdf (accessed on 10 February 2020).
54. Jönköping Municipality. Översiktplan 2015. 2015. Available online: https://www.jonkoping.se/byggabomiljo/planarbetestadsplaneringochutveckling/oversiktligplanering/gallandeoversiktsplaner/oversiktsplan2016.4.74fef9ab15548f0b8004c6.html (accessed on 7 February 2020).
55. City of Lindau. Climate Friendly Mobility Concept for Lindau (Klimafreundliches Lindauer Mobilitätskonzept). 2017. Available online: https://www.gtl-lindau.de/images/pdf_klimo/2017-06-21-Lindau%20-%20KLiMo_Endbericht (accessed on 10 February 2020).
56. City of Lindau. Local land use zoning plan and supporting documents (Flächennutzplan). 2013. Available online: https://www.stadtlindau.de/B%C3%BCrger-Politik-Verwaltung/Planen-Bauen/Fl%C3%A4chennutzungsplan (accessed on 10 February 2020).
57. City of Lindau. Integrated City Development Concept (ICDC). 2015. Available online: https://www.stadtlindau.de/B%C3%BCrger-Politik-Verwaltung/Planen-Bauen/Integiertes-Stadtentwicklungskonzept-ISEK- (accessed on 10 February 2020).
58. Lund Municipality. LundaMaTs III. Strategy for a sustainable transport system in Lund Municipality. 2014. Available online: https://www.lund.se/globalassets/lund.se/traf_infra/lundamats/lm_iii_strat_sv_visa.pdf (accessed on 11 February 2020).
59. Lund Municipality. Lunds kommuns översiktsplan. 2018. Available online: https://www.lund.se/trafik--stadsplanering/oversiktsplan/ (accessed on 11 February 2020).
60. Malmö Municipality. Trafik- och mobilitetsplan för ett mer tillgängligt och hållbart Malmö. 2016. Available online: https://malmo.se/download/18.16ac037b154961d0287b3d8/1491303428445/MALM_TROMP_210x297mm_SE.pdf (accessed on 12 February 2020).
61. Malmö Municipality. Översiktsplan för Malmö. 2018. Available online: https://malmo.se/download/18.270ce2fa16316b5786c18924/1528181608562/%C3%96VERSIKTSPLAN+F%C3%96R+MALM%C3%96_antagen_31maj2018_lowres.pdf (accessed on 12 February 2020).
62. City of Nottingham. Local Transport Plan. 2011. Available online: https://www.nottinghamcity.gov.uk/transportstrategies (accessed on 13 February 2020).
63. City of Nottingham. Local Plan 2017. 2017. Available online: https://www.nottinghamcity.gov.uk/information-for-business/planning-and-building-control/planning-policy/the-local-plan-and-planning-policy/the-adopted-local-plan/#:~{}:text=In%20Nottingham%20the%20Local%20Plan,Nottinghamshire%20and%20Nottingham%20Waste%20Core (accessed on 15 February 2020).
64. City of Tilburg. Mobility Strategy Tilburg, Together towards Smart Sustainable Mobility in 2040 (Mobiliteitsaanpak Tilburg Samen op weg naar 2040—Tilburg slim en duurzaam op weg). 2017. Available online: https://www.tilburg.nl/actueel/gebiedsontwikkeling/mobiliteitsplan-2040/#:~{}:text=Mobiliteitsaanpak%20Tilburg%3A%20samen%20op%20weg,dragen%20aan%20een%20leefbare%20stad (accessed on 15 February 2020).
65. City of Tilburg. Further development of Mobility Strategy (Uitwerking Mobiliteitsaanpak Tilburg). 2017. Available online: https://www.tilburg.nl/fileadmin/files/actueel/MobiliteitsAgenda013__sept_2017_.pdf (accessed on 15 February 2020).
66. City of Tilburg. Land Use Vision (Omgevingsvisie Tilburg). 2015. Available online: https://www.tilburg.nl/ondernemers/bouwen-en-vestigen/bestemmingsplan/ (accessed on 15 February 2020).
67. Ryan, G.W.; Bernard, H.R. Techniques to identify themes. *Field Methods* **2003**, *15*, 85–109. [CrossRef]

68. Patton, M.Q. Enhancing the quality and credibility of qualitative analysis. *Health Serv. Res.* **1999**, *34*, 1189–1208.
69. May, A.; Boehler-Baedeker, S.; Delgado, L.; Durlin, T.; Enache, M.; van der Pas, J.-W. Appropriate national policy frameworks for sustainable urban mobility plans. *Eur. Transp. Res. Rev.* **2017**, *9*, 7. [CrossRef]
70. Knieling, J.; Othengrafen, F. Planning culture—A concept to explain the evolution of planning policies and processes in Europe? *Eur. Plan. Stud.* **2015**, *23*, 2133–2147. [CrossRef]
71. Paez, A.; Scott, D.M.; Morency, C. Measuring accessibility: Positive and normative implementations of various accessibility indicators. *J. Transp. Geogr.* **2012**, *25*, 141–153. [CrossRef]
72. Boisjoly, G.; El-Geneidy, A.M. How to get there? A critical assessment of accessibility objectives and indicators in metropolitan transportation plans. *Transp. Policy* **2017**, *55*, 38–50. [CrossRef]
73. Akyelken, N.; Banister, D.; Givoni, M. The sustainability of shared mobility in London: The dilemma for governance. *Sustainability* **2018**, *10*, 420. [CrossRef]
74. Van Wee, B.; Handy, S. Key research themes on urban space, scale, and sustainable urban mobility. *Int. J. Sustain. Transp.* **2014**, *10*, 18–24. [CrossRef]
75. Garau, C.; Masala, F.; Pinna, F. Benchmarking smart urban mobility: A study on Italian cities. In *International Conference on Computational Science and Its Applications*; Springer: Cham, Switzerland, June 2015; pp. 612–623.
76. López-Lambas, M.E.; Corazza, M.V.; Monzón, A.; Musso, A. Rebalancing urban mobility: A tale of four cities. *Proc. Inst. Civ. Eng.—Urban Des. Plan.* **2013**, *166*, 274–287. [CrossRef]
77. Mozos-Blanco, M.A.; Menéndez, E.P.; Ruiz, R.M.A.; Baucells-Aletà, N. The way to sustainable mobility. A comparative analysis of sustainable mobility plans in Spain. *Transp. Policy* **2018**, *72*, 45–54. [CrossRef]
78. Kenworthy, J. Is automobile dependence in emerging cities an irresistible force? Perspectives from São Paulo, Taipei, Prague, Mumbai, Shanghai, Beijing, and Guangzhou. *Sustainability* **2017**, *9*, 1953. [CrossRef]

Article

# Concepts of Development of Alternative Travel in Autonomous Cars

**Vytautas Palevičius, Rasa Ušpalytė-Vitkūnienė, Jonas Damidavičius * and Tomas Karpavičius**

Department of Roads, Faculty of Environmental Engineering, Vilnius Gediminas Technical University, Saulėtekio al. 11, LT-10223 Vilnius, Lithuania; vytautas.palevicius@vgtu.lt (V.P.); rasa.uspalyte-vitkuniene@vgtu.lt (R.U.-V.); karpavicius.tomas@yahoo.com (T.K.)

* Correspondence: jonas.damidavicius@vgtu.lt; Tel.: +370-6-8553836

Received: 29 September 2020; Accepted: 21 October 2020; Published: 24 October 2020

**Abstract:** Autonomous car travel planning is increasingly gaining attention from scientists and professionals, who are addressing the integration of autonomous cars into the general urban transportation system. Autonomous car travel planning depends on the transport system infrastructure, the dynamic data, and their quality. The efficient development of travel depends on the development level of the Intelligent Transport Systems (ITS) and the Cooperative Intelligent Transport Systems (C-ITS). Today, most cities around the world are competing with each other to become the smartest cities possible, using and integrating the most advanced ITS and C-ITS that are available. It is clear that ITS and C-ITS are occupying an increasing share of urban transport infrastructure, so the complex challenges of ITS and C-ITS development will inevitably need to be addressed, in the near future, by integrating them into the overall urban transport system. With this in mind, the authors proposed three autonomous car travel development concepts that should become a conceptual tool in the development of ITS and C-ITS.

**Keywords:** development; autonomous cars; ITS and C-ITS; public infrastructure; Kendall method

## 1. Introduction

Every day, efforts are made to improve our environment and make it synonymous with a comfortable life; new products and technologies are being developed that are rapidly changing people's environments and movement habits. Famous car companies are competing with each other in an effort to bring widespread autonomous cars to the market by 2025. Publicly available information is not very detailed, despite many autonomous car surveys being already underway, but they usually cover one specific area only.

Car security systems have evolved very rapidly over the past few decades. All new engineering solutions have been developed to help the driver. These technologies help to prevent or reduce the consequences of accidents, but they do not guarantee complete safety. However, according to the United States Department of Transportation, 94 ± 2.2% of accidents are due to human error [1]. Researchers Bertoncello and Wee claim that autonomous car technology, used en masse, can reduce the number of accidents by as much as 90% [2], also supported by researchers in other countries that completed various studies, examining the key parameters that affect autonomous car safety: human-car interaction, cyber security, testing, societal adaptation, and others [3,4].

Autonomous cars entering the market will further increase the production volume of car manufacturers, which will be challenged by total person kilometers traveled by car, because people of all cultures and social groups will no longer have the obstacle of having to manually drive. Children, people with disabilities, and the elderly will have much easier access to transport, without having to drive. Replacing 50% of all vehicles in traffic with autonomous ones could save 9600 lives, reduce accidents by nearly 2 million, save 160 billion USD, reduce traffic congestion by 35 percent, and save about 1700 travel hours [5,6].

A Global Positioning System (GPS) or inertial measurement units (IMU) system can determine the exact position of a car, not only when it has a continuous flow of data, but also when the data are received with certain errors. Therefore, most experts or transport engineers and scientists working in the area of autonomous vehicles propose to connect both these systems so that there would be a possibility to precisely determine the positioning of the car in real time [7–9].

It is not just the connection between the car systems themselves that is important (V2V—Vehicle to Vehicle), but also the communication between car and road infrastructure (V2I—Vehicle to Infrastructure). This increases the responsibility of the infrastructure for the overall coordination of traffic flow and provides easier spatial orientation of the car with faster decision-making. This would have a major impact on road conductivity, car emissions, and time costs [10,11].

It is clear that the use of ITS in the city will be effective only if a unified system linking vehicles, multimodal transport infrastructure, and traffic management is developed. Data and information should be transferred from one vehicle to another (V2V), between vehicles and infrastructure (V2I), and vice versa, and from one piece of infrastructure to another (I2I). Cooperative Intelligent Transport Systems (C−ITS) enables real-time communication between vehicles, road infrastructure, and road users. The vehicles and road elements operating in the system have Intelligent Transport Systems (ITS) stations with a standardized structure, consisting of communication and hardware. Continuous wireless communication between all three groups ensures traffic safety, traffic management, efficiency, and generates a net positive impact on the environment. The main difference between C-ITS and ITS is that there is a connection between infrastructure and personal devices (V2P), and a connection between personal means and vehicles (P2V). Taking into account the available data, intelligent transport systems have been installed and developed, as well as the adaption of autonomous cars at the appropriate level in the common transport system.

Researchers in France and the United States point out, in an article published in 2016 [12], that the emergence of autonomous cars on city streets will pose enormous challenges. They agree that autonomous vehicles should reduce accidents, but will sometimes have to choose between two evils, such as driving through pedestrians or sacrificing oneself and/or one's passenger to save pedestrians.

Australian scientists proposed that shared autonomous vehicles (SAVs) could provide inexpensive mobility and on-demand services [13]. In addition, the autonomous vehicle technology could facilitate the implementation of dynamic ride-sharing (DRS). For this purpose, a stated choice survey was conducted and analyzed using a mixed logit model. The results show that service attributes, including travel cost, travel time, and waiting time, may be critical determinants of the use of SAVs and the acceptance of DRS. Also, the results imply that the adoption of SAVs may differ across cohorts, whereby young individuals and individuals with multimodal travel patterns may be more likely to adopt SAVs [13].

Ratner points out that cities and states must shape tax policy to directly confront the range of impacts autonomous vehicles will have over the next several decades. Autonomous vehicles are expected to decrease emissions overall and strengthen productivity [14].

Talebpour and Mahmassani focused on one of the most hazardous traffic situations: the possible collision between a pedestrian and a turning vehicle at signalized intersections. They presented a probabilistic model of pedestrian behavior to signalized crosswalks. In order to model the behavior of pedestrians, they took not only pedestrian physical states but also contextual information into account. They proposed a model based on the Dynamic Bayesian Network which integrated relationships among the intersection context information and the pedestrian behavior in the same way as a human. The particle filter was used to estimate the pedestrian states, including position, crossing decision, and motion type. Experimental evaluation using real traffic data showed that this model is able to recognize the pedestrian crossing decision in a few seconds from the traffic signal and pedestrian position information. This information was assumed to be obtained with the development of Connected Vehicle [15].

Millard-Ball performed an analysis without considering strategic interactions with other road users. He used game theory to analyze the interaction between pedestrians and autonomous vehicles, focusing on driving at a crossing. He found that the strategic shortcomings of the autonomous vehicle in cities would slow down all traffic [16].

Chinese and Pakistani researchers proposed a new cellular automata model to address this issue, where different autonomous and manual vehicles and manual are compared in terms of fundamental traffic parameters. Their model results suggest that autonomous vehicles can raise the flow rate of any network considerably, despite the running heterogeneous traffic flow [17].

Medina-Tapia and Robusté created a model for a circular city based on continuous approximations, considering demand surfaces over the city. Numerical results from our model predict direct and indirect effects of connected and autonomous vehicles. Direct effects will be positive for our cities: less street supply is needed to accommodate the traffic, congestion levels decrease: travel costs may decrease by 30%. Some indirect effects will counterbalance these positive effects: a decrease of 20% in the value of travel time can reduce the total cost by a third; induced demand could be as high as 50%, bringing equivalent total costs in the future scenario; the vehicle-kilometers traveled could also affect the future scenario; and increases in city size and urban sprawl [18].

A paper from Hungarian researchers proposed a tunable intersection control algorithm for autonomous vehicles, with the aim to minimize both traveling time and energy consumption while guaranteeing collision-free passage of the vehicles. For this purpose, a Model Predictive Control has been designed with tunable performances. The proposed algorithm was validated in the CarSim environment by comparing its solution with a simulation-based optimization. Also, a complex example was tested with a newly entering vehicle joining the optimization process. The results demonstrated the effectiveness of the proposed centralized controller [19].

Other researchers proposed a multilevel cloud system for autonomous vehicles which was built over the Tactile Internet. In addition, base stations at the edge of the radio-access network (RAN) with different technologies of antennas are used in our system. Finally, simulation results show that the proposed system with multilevel clouding can significantly reduce the round-trip latency and the network congestion. In addition, our system can be adapted in the mobility scenario [20].

An analysis of the scientific literature led to the conclusion that the breakthrough in the development of autonomous cars will only intensify in the near future. This breakthrough is likely to be further exacerbated by the European Union's green course, which is Europe's new growth strategy. Its goal is to become the world's first climate-neutral continent by 2050. All industrial value chains will play a key role in this process, as all investment will focus on increasing the economy's resilience to climate change. In order to contribute to the set goals, we, set the following tasks:

1. Identify groups of key criteria that will pose the greatest challenges for the development of autonomous cars in the city;
2. Carry out a survey of experts from both Lithuania and other countries (Sweden, Italy, Estonia, Latvia, Poland, and Croatia) and use the Kendall method to calculate the weights of the main groups of criteria;
3. Offer at least three autonomous car travel development concepts, which should become a conceptual tool for the ongoing development of ITS and C-ITS;
4. Present the conclusions of the scientific discussion at the end of the article.

## 2. Compilation and Evaluation of the List of Main Groups of Criteria

The list of the main groups of criteria was compiled by means of an expert analysis of various articles, published by researchers from other countries. After the authors agreed on the final list of criteria, it was submitted to the experts for evaluation. The experts were selected on the basis of work experience and possession of a scientific degree in the field of examination. The expert group consisted of 13 experts working in Lithuania (innovation, ITS, electronic communications, autonomous car legislation, and policy-making areas) and 8 experts from other countries (Sweden, Italy, Estonia, Latvia, Poland, and Croatia). Twenty-two experts questionnaires were sent for peer review, indicating the criteria of the main groups selected by the authors. The Kendall method was chosen for the evaluation, the essence of which is to evaluate the criteria on a scale of one to eight and to rank them from the highest to the lowest value of the criterion. Also, each expert was given the opportunity to comment and make suggestions on the selected criteria. Following the evaluation of the expert surveys, no comments or suggestions were received on the proposed criteria. After evaluating the questionnaires completed by twenty-two experts, twenty-one expert questionnaires were selected for further evaluation, but one questionnaire was incorrectly completed, and thus was rejected.

It is important to mention that the aim of our research was to identify the groups of key criteria that will pose the greatest challenges for the future development of autonomous cars in the city. Once the groups of key criteria have been identified, future scientific articles will examine the quantitative and qualitative criteria for each group separately. The following are selected key groups of criteria and their brief descriptions (Figure 1).

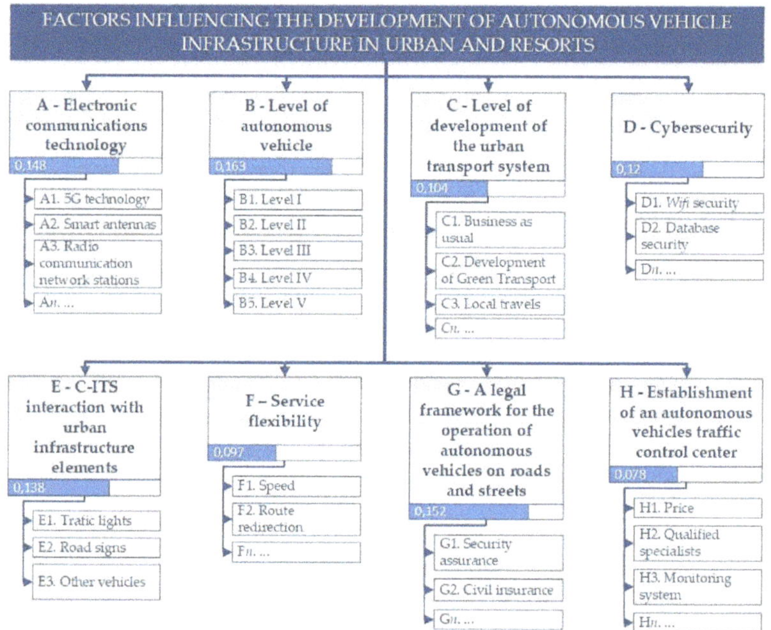


**Figure 1.** Criteria weights for the main groups of criteria.

**Group A.** Electronic communications technology. Recently, especially in countries in the European Union, public reactions and spreading misinformation on the topic of 5G communication have been observed. This group was selected by our co-authors because autonomous car manufacturers emphasize that 5G communication technologies, due to their features and essential operating principles, must emerge in cities to ensure the uninterrupted movement of autonomous cars. Also, we see a high risk of public concern about the dissemination of misinformation about the effects of 5G on the environment, human health, and even the links with COVID-19. However, the European Commission notes that the European Union has the highest standards of consumer protection in the world, the protection of public health and the security of citizens is a priority, and that the same highest standards of security apply to the new e-communication technology generation.

**Group B**. Level of car autonomy. Currently, most experts agree that the level of autonomy of the car is at technological level 3. According to the SAE International Technical Standard [21], a fully autonomous car is currently what is being pursued, with development moving in a consistent direction to reach such a goal. In addition, there are a variety of automated control options that are already available on the market that can automate part of the driving process. SAE International Technical Standard [21] provides a taxonomy for motor vehicle automation ranging in level from no automation to full automation (Table 1). However, it provides detailed definitions only for the highest three levels of automation provided in the taxonomy (namely, conditional, high, and full automation) in the context of motor vehicles (hereafter also referred to as "vehicle" or "vehicles") and their operation on public roadways. These latter levels of advanced automation refer to cases in which the dynamic driving task is performed entirely by an automated driving system during a given driving mode or trip. Popular media and legislative references to "autonomous" or "self-driving" vehicles encompass some or all of these levels of automation. These definitions can be used to describe the automation of (1) on-road vehicles, (2) particular systems within those vehicles, and (3) the operation of those vehicles. "On-road" refers to public roadways that collectively serve users of vehicles of all classes and automation levels (including no automation), as well as motorcyclists, pedal cyclists, and pedestrians.

**Table 1.** SAE International Technical Standard provides Terminology for Motor Vehicle Automated Driving Systems [21].

| SAE Level | Descriptive Name | Functional Description |
|---|---|---|
| 0 | Without autonomy | Only the driver controls all car systems |
| 1 | Driver's assistance | Very little help for the driver. For example: maintaining speed and distance according to the car in front and the surrounding environment; automatic light switching. |
| 2 | Partial autonomy | Assistance to the driver when the system can maintain speed, safety distance, and proper driving trajectory, taking into account the surrounding environment. |
| 3 | Conditional autonomy | Takeover of driving, provided that the driver assumes control at any time at the request of the system. The driver must always remain alert and monitor the situation. |
| 4 | High autonomy | Safe takeover of driving control, provided that even if there is a need to transfer control to the driver, the car itself will continue to move safely. |
| 5 | Full Autonomation | Safe and complete takeover of driving control, in all conditions the autonomous system safely copes with the route of a journey set by the human. |

The main higher goals of automatic steering are the following [22]:

- Safety: reduce accidents caused by human error;
- Efficiency and environmental objectives: increase the efficiency of the transport system and reduce congestion through new urban mobility solutions. It will also facilitate smoother urban traffic and reduce vehicle energy consumption and emissions;
- Convenience: enable user freedom for other activities when automated systems are activated;
- Social inclusion: ensure all user groups can utilize the transport, including the elderly and the disabled;
- Accessibility: facilitate access to the city center.

**Group C**. Level of development of the urban transport system. Each city's transport system and its development level are different and often depend on the historically formed network of streets and the location of buildings with various intended purposes. This set of criteria is very much needed for autonomous cars, which are assessed from levels 1–3 [21]. As a practical example, in 2017, for the first time in Lithuania, an autonomous vehicle was tested on the streets of Vilnius Old Town. During testing, the vehicle faced challenges with the level of infrastructure development of the existing urban transport system. The autonomous car was not able to recognize the stationary "green arrow sign" on the traffic lights, the intensive movement of pedestrians and cyclists in the narrow streets of the old town hindered the smooth movement of the autonomous vehicle, and it was unable to recognize temporary road signs, and so on.

**Group D.** Cyber security. It is obvious that the technologies of the future will radically change our lives—autonomous smart devices (pumps, lawn mowers, etc.), autonomous cars, virtual reality, and so on. Some experts estimate that there are currently about 200 billion devices connected to the Internet that are programmable and automated, and such changes are drastically changing our lifestyle habits. However, there is a huge problem—most technologies can be hacked or disabled. Every year, thousands of security vulnerabilities are discovered in the equipment of well-known manufacturers. A few of them are some of the strongest in cyber security, but still face a myriad of security challenges every year.

**Group E.** Interoperability of C-ITS with urban infrastructure elements. This set of criteria is one of the most important for the interaction of autonomous vehicles with the ecosystem of infrastructure elements. In order for autonomous vehicles to move freely on city streets, safe and developed infrastructure must be in place, and to achieve this, communication between vehicles and infrastructure must be also be present. The importance of this set of criteria can be justified by a South Korean research project that, in 2017, installed a 320,000 $m^2$ city-wide area (K-City) for autonomous cars that is being tested under real conditions.

**Group F.** Service flexibility. It is likely that autonomous cars will appear in every city in the near future, although in the beginning, they may be used mainly in public transport, designed to carry a larger number of passengers. However, for them to appear on urban streets, a competitive range of services must be offered to compete with what traditional public transport has to offer. For example, autonomous cars must be fast, safe, comfortable, have the ability to re-route in the event of a change in weather conditions or accidents, and overall offer a more effective service.

**Group D.** A legal framework has been created that does not hinder the operation of autonomous vehicles on roads and streets. Most countries in the world are already working intensively on the legal framework for autonomous vehicles. The aim is to allow manufacturers not only to test autonomous vehicles as soon as possible, but also to allow consumers to use them safely in public traffic. For example, in 2011, the U.S. state of Nevada was the first to legalize autonomous car testing on public roads and Japan launched the first autonomous car tests on public roads in 2013. EU countries are also working hard on changes to the legislation, but there is a greater focus on autonomous car testing. The reason for this is that it is not yet entirely clear who takes responsibility for accidents in public traffic; when the accident occurred through the fault of the autonomous vehicle, the conditions of compulsory civil insurance, etc., have not been resolved and discussed.

**Group H.** Establishment of an autonomous traffic management center. As the number of autonomous cars will only increase in the future and they will be moving without drivers, there will be a need to set up autonomous traffic control centers. Universities will have to develop study programs that will train suitably qualified IT specialists, autonomous vehicle operators, autonomous transport engineers, and so on. These professionals will need to be able to remotely implement coordination processes related to the autonomous vehicle infrastructure ecosystem. The main function of this center would be to provide autonomous vehicle management services, autonomous vehicle ecosystem monitoring, and other functions.

Following an expert survey, the consistency of expert opinions needs to be assessed. When calculating the correlation coefficient, it is possible to determine whether the opinions of two experts are agreed, but if the number of experts is higher, the concordance coefficient W is calculated according to the concordance coefficient W [23–27].

First, the sum of the rank of each evaluation indicator is determined according to Equation (1):

$$P_j = \sum_{k=1}^{r} P_{jk}, \tag{1}$$

where $P_{jk}$—$k$-th expert evaluation for the $j$-th assessment indicator, $k$—the experts authority coefficient.

The following is the average value of the evaluation indicator according to Equation (2):

$$\overline{P_j} = \frac{\sum_{k=1}^{r} P_{jk}}{r}, \tag{2}$$

where "$r$" is the number of experts.

The sum of the squares of the deviations of the evaluation results for each evaluation indicator according to Equation (3):

$$S = \sum_{j=1}^{m} (P_j - \overline{P})^2, \tag{3}$$

where "$m$" is the number of evaluation indicators.

The average sum of the ranks of all evaluation indicators is calculated according to Equation (4):

$$\overline{P} = \frac{\sum_{k=1}^{r} \sum_{j=1}^{m} P_{jk}}{m}. \tag{4}$$

The concordance coefficient $W$ is calculated according to Equation (5):

$$\chi^2 = W \cdot r \cdot (m-1) = \frac{12S}{r \cdot m \cdot (m+1)}. \tag{5}$$

If the $\chi^2$ calculated using Equation (5) is greater than the $\chi^2{}_{lent}$ according to the table, obtained from the table expression $\chi^2$ table with a certain degree of freedom $v = m - 1$, and assuming a significance level of $\alpha = 0.05$ (the usual value of $\alpha$ used in practice), the expert opinions are considered to be consistent.

When calculating the significance of evaluation indicators, the best value is the largest numerical expression [28–30]. The significance of the evaluation indicators is calculated according to Equation (6):

$$\overline{q}_j = \frac{\overline{P}_j}{\sum_{i=1}^{m} P_j}. \tag{6}$$

The sum of the significance indicators of the calculated assessment indicators is equal to 1. Significance indicates how many times one assessment indicator is more important than another assessment indicator. Taking advantage of this, it is possible to carry out a further selection of assessment indicators and select only those assessment indicators that have the highest value in each intended group [31–34].

Calculations using the Kendall method show that the group B criterion, "level of car autonomy" (0.163), will have the greatest weight for the development of alternative car trips with autonomous cars. The second most important was elected by the experts to be the Group G criterion "legal framework created that does not hinder the operation of autonomous vehicles on roads and streets" (0.152). The experts chose third place to be Group A criterion "electronic communications technology" (0.148). More detailed criteria weights for the main groups of criteria are shown in Figure 1.

## 3. Autonomous Car Travel Development Concepts

Taking into account the available data, intelligent transport systems have been installed and developed, as well as the adaption of autonomous cars at the appropriate level in the common transport system, the authors have offered three autonomous car trip development concepts.

The first concept involves autonomous smart travel planning from the point of choice to the final point (Figure 2). Travel planning is performed using only developed intelligent transport systems and the start and end points are determined using GPS coordinates. According to the obtained GPS movement, vehicles, and services are selected for possible use (private car, public transport (buses, trolleybuses, metro, tram, etc.)), as well as transport sharing services (cars, bicycles, scooters, etc.), and transportation services (on-demand transport) or a combination thereof, and the trip route is planned and finalized. The travel plan also includes different types of movement (by foot, while driving or waiting), as well as providing vehicle (mode of transportation) change points (stops, stations, Park and Ride sites, etc.).

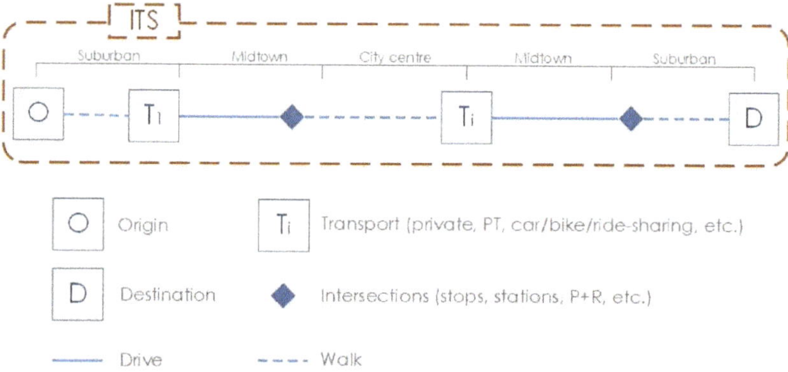


**Figure 2.** The first concept.

Prerequisites for ITS to operate throughout the whole territory, i.e., in the suburbs, the midtown area, and the city center, and the location of public transport stops, transfer and traffic sharing service sites must be close to the start and end points.

This concept could include level 0–1 autonomous vehicles, where the driver must monitor traffic conditions and control the vehicle during the journey. In essence, this concept does not replace the existing transport system, it facilitates only the planning and re-planning of the travel route in the current time, depending on the traffic situation and the circumstances prevailing throughout the transport system. Thanks to convenient travel planning and a large choice of vehicles, this concept allows you to partially give up your own car, requires no greater need for parking spaces in the city, and reduces vehicle traffic in the central area of the city.

The second concept involves self-intelligent travel planning from the point of choice to the end point, involving autonomous vehicles operating in the transport system (Figure 3). Travel planning is carried out using both developed intelligent transport systems and adapting the capabilities of cooperative intelligent transport systems. The start and end points are determined using GPS coordinates. According to the obtained trajectory, possible vehicles and services are selected (analogous to the first concept) or a combination thereof, and autonomous vehicles or their systems (public transport, transport sharing services) are included and the trip is planned. The travel plan also includes different types of movement (by foot, while driving or waiting), as well as providing vehicle (mode of transportation) change points (stops, stations, Park and Ride sites, etc.).

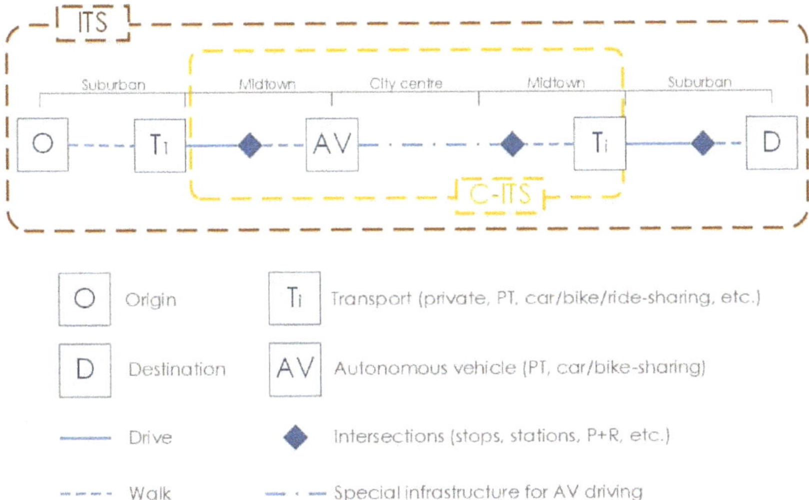


**Figure 3.** The second concept.

According to this concept, ITS should operate throughout the territory, i.e., in the suburbs, the midtown area, and the city center, and the operation of C-ITS should be ensured in the central part of the city, and as far as possible into the midtown area. Such a communication system could be operated by level 2–3 autonomous vehicles, where the AV accelerates, brakes, and drives itself, but the driver must be ready to drive the car in case of changes in traffic conditions and take control of the vehicle at any time. In areas with an uninterrupted traffic regime (often in the midtown part of the city), the driving environment is monitored by the autonomous car system, in areas with heavy traffic with different transport modes (often in the city center, where it is difficult to predict obstacles—pedestrians, cyclists, goods delivery transport, etc.—the driving environment must be monitored by the driver and, if necessary, the driver should take control of the autonomous car).

The operation of this concept requires the development of additional infrastructure to ensure a smooth real-time connection between vehicles, road infrastructure, and road users. The safe and secure operation of the system requires a communication signal that is reliable, free, and has a low latency. The main means used to support such a communication signal are Dedicated Short Range Communications (DSRC). This connection uses 5.9 GHz radio waves. It is very important that C-ITS can interact not only with the 5.9 GHz signal, but also with other rapidly evolving signals, such as 2G, 3G, 4G, WiMax, and so on. Also, due to the different operating principles of AV and conventional cars, they need to provide a separate traffic infrastructure or redevelop the existing, separate lanes only for AV.

Thanks to convenient travel planning and a large selection of vehicles, this concept allows you to partially give up your own car. The operation of AV in the central part of the city would reduce investments in transport infrastructure (parking lots, new street connections) by reducing the number of vehicles, reducing traffic congestion, increasing the attractiveness of autonomous public transport or transport sharing services.

The third concept is described as including intelligent smart travel planning from the point of choice to the final point, using only autonomous vehicles (Figure 4). Travel planning is carried out using both developed intelligent transport systems and adapting the capabilities of cooperative intelligent transport systems. The start and end points are determined using GPS coordinates. Based on the obtained trajectory, the AV itself selects the optimal route and plans the trip. In this concept, AV can be used as any mode of transport (private transport, public transport, transport sharing services, freight transport, shuttle/taxi services, etc.). The travel plan also includes different types of movement (by foot, while driving or waiting), as well as providing vehicle (mode of transportation) change points (stops, stations etc.).

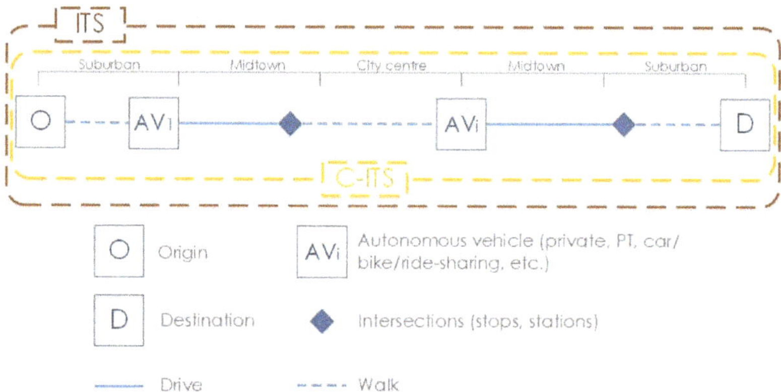


**Figure 4.** The third concept.

According to this concept, ITS, and as far as possible C-ITS, must operate throughout the territory, i.e., in the suburbs, the midtown area, and the city center. Such a communication system could operate level 4–5 autonomous vehicles, where AV automatic control can fully control the vehicle, maintain vehicle control, even when the driver does not intervene in the event of changed driving conditions (for example, in an emergency).

The existing communication infrastructure can be adapted to the operation of this concept; it is necessary only to ensure the communication signals for the operation of the system, which would be reliable, free of charge, and with a short time delay.

This concept allows you to give up your own car, but there is a certain need to have your own AV. In any case, heavy AV traffic would be more efficient than any road transport system in operation to date. The integration of AV into the transport system would reduce investments in transport infrastructure (parking lots, new street connections, etc.), significantly reduce air pollution, the subsequent use of resulting free areas for green spaces, investment projects, and reduce traffic congestion and the number of accidents.

The proposed concepts for the development of autonomous cars infrastructure are principled, which should be used as a core for the development of further details of the systems. It is very important to understand that each city's transport system is distinctive and has some unique features, while cities themselves and their urban development can also influence the application of the proposed concepts. Therefore, the concepts proposed by the authors should be seen as an axis for infrastructure development, while the specific elements, the combination of interoperability between the systems, etc., should be analysed in the context of a particular city.

## 4. Conclusions

Calculations using the Kendall method show that the criteria of the "car automation level" group (0.163), "created legal basis that does not interfere with the operation of autonomous vehicles on roads and streets" group (0.152), and "electronic communications technology" group criteria (0.148) will have the greatest weight for the development of alternative autonomous trips.

Taking into account the available data, intelligent transport systems have been installed and developed, as well as the adaptation of autonomous cars at the appropriate level in the common transport system, the best three autonomous car trip development concepts are as follows:

- The first development option includes autonomous smart travel planning from the point of choice to the final point (travel planning is performed using only developed intelligent transport systems and possible vehicles and services or a combination thereof are selected according to the obtained trajectory, after which the trip is planned);
- The second development option is where autonomous intelligent travel planning from the point of choice to the final point, and includes autonomous vehicles operating in the transport system (travel planning is carried out using both developed intelligent transport systems and adapting the capabilities of cooperative intelligent transport systems possibilities, as well as possible vehicles and services, which are selected and the trip is planned according to the obtained trajectory);
- The third development option is where independent intelligent travel planning from the point of choice to the final point is performed by using only autonomous vehicles (travel planning is performed using both developed intelligent transport systems and the capabilities of cooperative intelligent transport systems, as well as the AV, which itself chooses the optimal route and plans a trip).

Of all the possible development options, the first would have the least impact on the existing urban transport system, while the most realistic development option would be the second, where travel planning involves both autonomous and other vehicles. The second development option could be a step towards the achievement of third, where all travel planning is done using only autonomous vehicles.

**Author Contributions:** All authors contributed equally to this work. V.P. developed project idea, analysed data, conducted the expert interviews, contributed to draft the paper. J.D. developed the model of quantitative and qualitative data collection, created graphs and figures, outlined conclusions and recommendations, performed calculations of weights and multiple criteria analysis, created tables, and contributed to draft the paper. R.U.-V. developed project idea, led the development of the methodology and contributed to draft the paper. T.K. analysed data, contributed to the revision of draft version and improvement of final paper. All authors discussed the results and commented on the paper. All authors have read and agreed to the published version of the manuscript.

**Funding:** This research was funded by Vilnius Gediminas Technical University.

**Conflicts of Interest:** The authors declare no conflict of interest.

## References

1. U.S. Department of Transportation National Highway Traffic Safety Administration. Traffic Safety Facts. 2015. Available online: https://crashstats.nhtsa.dot.gov/Api/Public/ViewPublication/812115 (accessed on 13 August 2020).
2. Bertoncello, M.; Wee, D. Ten Ways Autonomous Driving Could Redefine the Automotive World. 2015. Available online: http://www.mckinsey.com/insights/automotive_and_assembly/ten_ways_autonomous_driving_could_redefine_the_automotive_world (accessed on 12 November 2012).
3. Tokody, D.; Albini, A.; Ady, L.; Raynai, Z.; Pongrácz, F. Safety and Security through the Design of Autonomous Intelligent Vehicle Systems and Intelligent Infrastructure in the Smart City. *Interdiscip. Descr. Complex Syst.* **2018**, *16*, 384–396. [CrossRef]
4. Koopman, P.; Wagner, M. Autonomous Vehicle Safety: An Interdisciplinary Challenge. *IEEE Intell. Transp. Syst. Mag.* **2017**, *9*, 90–96. [CrossRef]
5. Thierer, A.D.; Hagemann, R. Removing Roadblocks to Intelligent Vehicles and Driverless Cars. *Wake For. JL Pol'y* **2014**, *5*, 339. [CrossRef]
6. Young, K.L.; Rudin-Brown, C.M. Designing Automotive Technology for Cross-Cultural Acceptance. In *Driver Acceptance of New Technology*; Ashgate Publishing Limited: Surrey UK, 2018; pp. 317–332.
7. Wang, S.; Deng, Z.; Yin, G. An Accurate GPS-IMU/DR Data Fusion Method for Driverless Car Based on a Set of Predictive Models and Grid Constraints. *Sensors* **2016**, *16*, 280. [CrossRef]
8. Dedes, G.C.; Mouskos, K.C. GPS/IMU/Video/Radar Absolute/Relative Positioning Communication/Computation Sensor Platform for Automotive Safety Applications. U.S. Patent 8,639,426, 28 January 2014.
9. Carratu, M.; Iacono, S.D.; Pietrosanto, A.; Paciello, V. Self-alignment procedure for IMU in automotive context. In Proceedings of the 2019 IEEE 17th International Conference on Industrial Informatics (INDIN), Helsinki, Finland, 22–25 July 2019; Volume 1, pp. 1815–1820.

10. Malinverno, M.; Avino, G.; Casetti, C.; Chiasserini, C.F.; Malandrino, F.; Scarpina, S. Performance Analysis of C-V2I-Based Automotive Collision Avoidance. In Proceedings of the 2018 IEEE 19th International Symposium on "A World of Wireless, Mobile and Multimedia Networks" (WoWMoM), Chania, Greece, 12–15 June 2018; pp. 1–9. [CrossRef]
11. Dong, C.; Wang, H.; Li, Y.; Liu, Y.; Chen, Q. Economic comparison between vehicle-to-vehicle (V2V) and vehicle-to-infrastructure (V2I) at freeway on-ramps based on microscopic simulations. *IET Intell. Transp. Syst.* **2019**, *13*, 1726–1735. [CrossRef]
12. Bonnefon, J.-F.; Shariff, A.; Rahwan, I. The social dilemma of autonomous vehicles. *Science* **2016**, *352*, 1573–1576. [CrossRef]
13. Haboucha, C.J.; Ishaq, R.; Shiftan, Y. User preferences regarding autonomous vehicles. *Transp. Res. Part C Emerg. Technol.* **2017**, *78*, 37–49. [CrossRef]
14. Ratner, S. Taxation of Autonomous Vehicles in Cities and States. 2018. Available online: https://papers.ssrn.com/sol3/papers.cfm?abstract_id=3285525 (accessed on 4 August 2020).
15. Talebpour, A.; Mahmassani, H.S. Influence of connected and autonomous vehicles on traffic flow stability and throughput. *Transp. Res. Part C Emerg. Technol.* **2016**, *71*, 143–163. [CrossRef]
16. Millard-Ball, A. Pedestrians, Autonomous Vehicles, and Cities. *J. Plan. Educ. Res.* **2016**, *38*, 6–12. [CrossRef]
17. Tanveer, M.; Kashmiri, F.A.; Naeem, H.; Yan, H.; Qi, X.; Rizvi, S.M.A.; Wang, T.; Lu, H. An Assessment of Age and Gender Characteristics of Mixed Traffic with Autonomous and Manual Vehicles: A Cellular Automata Approach. *Sustainability* **2020**, *12*, 2922. [CrossRef]
18. Medina-Tapia, M.; Robusté, F. Implementation of Connected and Autonomous Vehicles in Cities Could Have Neutral Effects on the Total Travel Time Costs: Modeling and Analysis for a Circular City. *Sustainability* **2019**, *11*, 482. [CrossRef]
19. Mihály, A.; Farkas, Z.; Gáspár, P. Multicriteria Autonomous Vehicle Control at Non-Signalized Intersections. *Appl. Sci.* **2020**, *10*, 7161. [CrossRef]
20. Khayyat, M.; Alshahrani, A.; Alharbi, S.; Elgendy, I.; Paramonov, A.; Koucheryavy, A. Multilevel Service-Provisioning-Based Autonomous Vehicle Applications. *Sustainability* **2020**, *12*, 2497. [CrossRef]
21. SAE International. SAE International Technical Standart Provides Terminology for Motor Vehicle Automated Driving Systems. 2014. Available online: http://standards.sae.org/j3016_201401/ (accessed on 4 August 2020).
22. ERTRAC. Connected Automated Driving Roadmap. Deliverable on ERTRAC Working Group "Connectivity and Automated Driving". 2019. Available online: https://www.ertrac.org/uploads/documentsearch/id57/ERTRAC-CAD-Roadmap-2019.pdf (accessed on 13 August 2020).
23. Kendall, M. *Rank Correlation Methods*, 4th ed.; Griffin: London, UK, 1970; 272p.
24. Kendall, M.G.; Gibbons, J.D. *Rank Correlation Methods*, 5th ed.; Oxford University Press: New York, NY, USA, 1990.
25. Zavadskas, E.K.; Peldschus, F.; Kaklauskas, A. *Multiple Criteria Evaluation of Projects in Construction*; Technika: Vilnius, Lithuania, 1994.
26. Zavadskas, E.K.; Vilutienė, T. A multiple criteria evaluation of multi-family apartment block's maintenance contractors: I—Model for maintenance contractor evaluation and the determination of its selection criteria. *Build. Environ.* **2006**, *41*, 621–632. [CrossRef]
27. Palevičius, V.; Burinskienė, M.; Antucheviciene, J.; Šaparauskas, J. Comparative Study of Urban Area Growth: Determining the Key Criteria of Inner Urban Development. *Symmetry* **2019**, *11*, 406. [CrossRef]
28. Ginevicius, R. A new determining method for the criteria weights in multi-criteria evaluation. *Int. J. Inf. Technol. Decis. Mak.* **2011**, *10*, 1067–1095. [CrossRef]
29. KrishanKumar, R.; Ravichandran, K.; Kar, S.; Cavallaro, F.; Zavadskas, E.K.; Mardani, A. Scientific Decision Framework for Evaluation of Renewable Energy Sources under Q-Rung Orthopair Fuzzy Set with Partially Known Weight Information. *Sustainability* **2019**, *11*, 4202. [CrossRef]
30. Roy, J.; Sharma, H.K.; Kar, S.; Zavadskas, E.K.; Saparauskas, J. An extended COPRAS model for multi-criteria decision-making problems and its application in web-based hotel evaluation and selection. *Econ. Res. -Ekon. Istraživanja* **2019**, *32*, 219–253. [CrossRef]
31. Lakusic, S.; Barauskas, A.; Mateckis, K.J.; Palevičius, V.; Antucheviciene, J. Ranking conceptual locations for a park-and-ride parking lot using EDAS method. *J. Croat. Assoc. Civ. Eng.* **2018**, *70*, 975–983. [CrossRef]

32. Podvezko, V.; Zavadskas, E.K.; Podviezko, A. An extension of the New Objective Weight Assessment Methods Cilos and Idocriw to Fuzzy Mcdm. *Econ. Comput. Econ. Cybern. Stud. Res.* **2020**, *54*, 59–76. [CrossRef]
33. KrishanKumar, R.; Mishra, A.R.; Ravichandran, K.S.; Peng, X.; Zavadskas, E.K.; Cavallaro, F.; Mardani, A. A Group Decision Framework for Renewable Energy Source Selection under Interval-Valued Probabilistic linguistic Term Set. *Energies* **2020**, *13*, 986. [CrossRef]
34. Zolfani, S.H.; Chatterjee, P. Comparative Evaluation of Sustainable Design Based on Step-Wise Weight Assessment Ratio Analysis (SWARA) and Best Worst Method (BWM) Methods: A Perspective on Household Furnishing Materials. *Symmetry* **2019**, *11*, 74. [CrossRef]


**Publisher's Note:** MDPI stays neutral with regard to jurisdictional claims in published maps and institutional affiliations.

*Article*

# Hub-Periphery Hierarchy in Bus Transportation Networks: Gini Coefficients and the Seoul Bus System

**Chansoo Kim [1,†], Segun Goh [2,†], Myeong Seon Choi [3], Keumsook Lee [4,*] and M. Y. Choi [5,*]**

1 AI Lab., Computational Science Center & ESRI, Korea Institute of Science and Technology, Seoul 02792, Korea; eau@econ.kist.re.kr

2 Institut für Theoretische Physik II, Heinrich-Heine-Universität Düsseldorf, 40225 Düsseldorf, Germany; segun.goh@gmail.com

3 Graduate School of Nanoscience and Technology, Korea Advanced Institute of Science and Technology, Daejeon 34141, Korea; joyms514@kaist.ac.kr

4 Department of Geography, Sungshin Women's University, Seoul 02844, Korea

5 Department of Physics and Astronomy and Center for Theoretical Physics, Seoul National University, Seoul 08826, Korea

* Correspondence: kslee@sungshin.ac.kr (K.L.); mychoi@snu.ac.kr (M.Y.C.)

† Equal contributions.

Received: 1 August, 2020; Accepted: 2 September 2020; Published: 6 September 2020

**Abstract:** Bus transportation networks are characteristically different from other mass transportation systems such as airline or subway networks, and thus the usual approach may not work properly. In this paper, to analyze the bus transportation network, we employ the Gini coefficient, which measures the disparity of weights of bus stops. Applied to the Seoul bus system specifically, the Gini coefficient allows us to classify nodes in the bus network into two distinct types: hub and peripheral nodes. We elucidate the structural properties of the two types in the years 2011 and 2013, and probe the evolution of each type over the two years. It is revealed that the hub type evolves according to the controlled growth process while the peripheral one, displaying a number of new constructions as well as sudden closings of bus stops, is not described by growth dynamics. The Gini coefficient thus provides a key mathematical criterion of decomposing the transportation network into a growing one and the other. It would also help policymakers to deal with the complexity of urban mobility and make more sustainable city planning.

**Keywords:** bus transportation system; urban infrastructure; network growth; Gini coefficient; complex systems

---

## 1. Introduction

Complex networks have attracted much attention in various research areas. Since the seminal works on small-world networks [1] and scale-free networks [2], interest in spatial networks has grown significantly including various fields of application [3,4]. Until recent years, there has been much effort to assess the sustainability of transportation networks [5–11]. In view of sustainability, a better understanding of the network structure is required for improving those analyses [12–15]. Specifically, there have appeared studies of mass transportation networks: For instance, the network structure and passenger flows of the Seoul subway system were analyzed [16,17]. Further, the growth process of the Seoul subway network was described by a master equation for a Yule-type model [18], and the passenger flow on the network was described by a gravity model modified by a Hill function [19]. Another study includes the clique growth model, in which the growth of the network is governed by attaching new cliques to the system [20]. In the case of the bus transportation network (BTN), scaling and renormalization ideas manifested the emergence of criticality in the passenger flow [21] while accessibility measurement was also successfully performed [22].

Among those spatial transportation networks explored, the BTN plays a substantial role in short- to mid-distance trips of passengers in a city. Therefore, changes in the BTN are rather important to network users (passengers) and network designers (policymakers). The BTN possesses some unique characteristics, different from those of other transportation systems [7,9,11,15,23]. Most of all, it serves periphery of a city with convenient and tangible transportation means to and from less noticed places in which other mass transportation systems such as the subway do not provide stops or stations [15]. Especially, in Seoul, while subway transportation has a major share for downtown areas including central business districts, residents in the suburbs take bus transportation for commuting. Part of the reason is that there are many mountainous and hilly areas in Seoul, which hinders the construction of subway stations at low cost [13]. In particular, the BTN changes, namely, vary in nodes (stops) and links (routes), usually much faster than others [24]. Furthermore, while a subway network has no noticeable hubs, the BTN has hubs and peripheral nodes that are distinct strikingly from each other.

Such characteristics make it rather inappropriate to apply existing models to the BTN [19–22]. To circumvent this difficulty, we in this paper propose a classification scheme to filter peripheral nodes out of the network with the aid of the disparity concept. The connectivity of each node, which is reflected in the weight, is compared; this manifests the duplex nature of the BTN and allows one to divide the nodes into two types, hubs and peripheries. We then probe the evolution of the two types from the viewpoint of the Yule process, which describes the well-known growth of an object growing proportionally to its current size. The Yule-type controlled growth process, extended and described generally by a master equation [25–27], was applied successfully to various growing complex systems [28,29]. It turns out that the growth process does not give a proper explanation for the evolution of the peripheral network displaying too many changes such as frequent constructions and closings of stops. Accordingly, it is indeed desirable to separate peripheral nodes from hubs, where the Gini coefficient [30,31] plays a key role. The Gini coefficient has long and widely been used in various fields [32,33]. There have been various interesting attempts to utilize it for separating data sets [34–36], recently including machine learning techniques [37,38].

The paper is organized as follows: Section 2 presents the Seoul bus transportation network, and we describe how one utilizes the Gini coefficient to assess the BTN. Each node in the BTN is categorized as a hub or a peripheral one, according to the Gini coefficient. The set of hub nodes composes the hub network, and in the same manner, the peripheral network is built of peripheral nodes. In Section 3, we probe how the hub network and the peripheral network changed over the two years, 2011 and 2013, and elucidate the evolution of the hub by means of the controlled growth mechanism. Section 4 summarizes our work and discusses implications for policymakers to overcome the vulnerability of the BTN as well as to improve it consistently.

## 2. Materials and Methods

As an exemplary BTN, we consider the Seoul bus system, which is a well-developed, large-scale BTN. We analyze the smart card (called *T-money card*) data, which are collected by the Seoul metropolitan government and not open to the public. The information contains the departure/arrival bus stops and times of each trip. Obtaining permission from the government, we have access to the dataset with the personal identification removed. Analyzing the passenger flow data for the years 2011 and 2013, we observe that the data as a whole resist explanation in terms of the controlled growth mechanism. We thus examine the hierarchical structure in the Seoul bus system by the help of the Gini coefficient, in advance of considering evolution as a controlled growth process.

### 2.1. Seoul Bus System

In the years 2011 and 2013, there were $N_s$ = 15,515 and 15,702 bus stops, respectively, in the Seoul bus system, the spatial distributions of which are shown in Figure 1. Among those, 8408 stops turn out to be common in both years, as recognized by their identifications (IDs). On April 11th in the year 2011, 5,668,431 people in total rode on buses and only $N_e$ = 12,908 stops out of the total have non-zero values of the strength. Here the strength $s$ of a stop stands for the total number of passengers departing/arriving that stop, i.e., getting on/off buses at the stop, while the weight $w_{ij}$ $(= w_{ji})$ between two stops $i$ and $j$ refers to the total number of passengers making trips between the two stops, i.e., getting on buses at stop $i$ and off at stop $j$ or vice versa. Accordingly, the weight $w_{ij}$ and the strength $s_i$ of stop $i$ are related via: $s_i = \sum_j w_{ij}$.

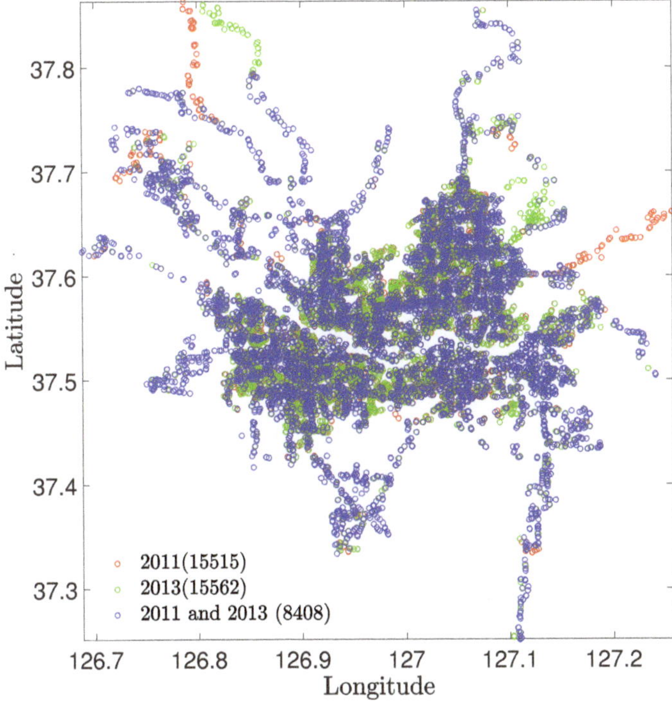


**Figure 1.** Locations of bus stops in the Metropolitan Seoul bus system: Blue dots represent bus stops persisting between the two years 2011 and 2013. Bus stops present only in 2011 and in 2013 are represented by red and green dots, respectively. Numbers in parentheses refer to the number of corresponding stops.

We first probe the growth of the Seoul BTN with the aid of the Yule-type growth model, which was successfully applied to the Seoul subway network to reveal the emergence of log-normal, Weibull, and power-law distributions [18]. In sharp contrast, neither strength distributions nor weight distributions of the Seoul BTN appear consistent with the Yule-type growth model.

For instance, Figure 2 presents distributions of the strength data on 11 April 2011 and on 4 March 2013, together with the corresponding log-normal distributions predicted by the Yule-type growth model. As shown, the strength distributions exhibit conspicuous deviations, for small strengths ($s \lesssim 100$), from the log-normal and Weibull distributions as well as the power-law distribution. Assuming that such disagreement results from the hierarchy among the stops, we attempt to overcome this issue by decomposing the bus stops into the two categories, hubs and peripheries, and considering the two separately. Namely, we presume that hubs and peripheries play different roles in the BTN. It is then plausible that the growth mechanism for these two types of nodes should be different from each other.

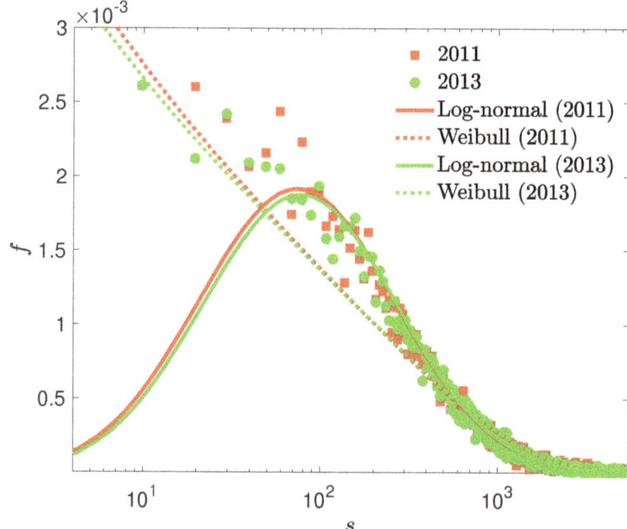


**Figure 2.** Distribution $f$ of strengths $s$ in the Seoul BTN, plotted on the semilogarithmic scale. Red squares and green circles are data points for the years 2011 and 2013, respectively. Solid and dotted lines plot the log-normal and the Weibull distributions, respectively, obtained via the maximum likelihood estimation from the data for $s \geq 25$. The location and the scale parameters of the log-normal distributions are given by $(\mu, \sigma) = (5.91, 1.27)$ for 2011 and $(5.92, 1.26)$ for 2013. The shape and the scale parameters of the Weibull distributions are estimated as $(\gamma, \eta) = (0.78, 700)$ for 2011 and $(0.79, 706)$ for 2013.

### *2.2. Gini Coefficients as a Classification Criterion*

The strength simply quantifies the overall amount of passenger flows of a bus stop, and there is no *a priori* reason to prefer a criterion for categorization from a strength profile, for example, by setting an ad-hoc threshold value. Moreover, the location-based approach does not seem to be proper for the bus network, because peripheral nodes can also be located not only at the end of a route but also in the center of the city. Here we thus proceed one step further to characterize the flow distribution of each bus stop connected to other stops in the network. Specifically, we propose the *Gini coefficient*, which measures how evenly the passenger flows are distributed from one stop to others, to provide a sensible criterion serving the above purpose. The coefficient has been widely used to quantify inequality of a given system in view of distributions [30,31]. Suppose that in a society consisting of $N\,(\gg 1)$ people, the $j$th person possesses wealth $x_j$. The Gini coefficient for the society is then defined to be

$$G \equiv \frac{\sum_{j=1}^{N} \sum_{k=1}^{N} \left| x_j - x_k \right|}{2N \sum_{j=1}^{N} x_j}, \tag{1}$$

which ranges from zero (completely equal) to unity (completely unequal): $0 \leq G \leq 1$. It focuses on the average absolute difference between all pairs in the society.

We now consider, in place of $x_j$, the weight $w_{ij}$ between given stop $i$ and other stops $j$ in the system. In principle, to reflect the connectivity structure of the BTN with meaningless pairs discarded, one needs to take into account bus routes serving the bus stop $i$. However, since buses always operate in one direction at each bus stop and bus routes may have complicated topology in general [22], this turns out to be a demanding task. Therefore, in this study, we simply adopt the weight as an index assessing the connectivity between bus stops. This leads to the generalized Gini coefficient $G_i$, defined for every stop $i$ in the BTN:

$$G_i \equiv \frac{\sum_{j=1}^{N_i} \sum_{k=1}^{N_i} |w_{ij} - w_{ik}|}{2N_i \sum_{j=1}^{N_i} w_{ij}}, \tag{2}$$

where $N_i$ denotes the number of bus stops with non-zero passenger flows to and from the $i$th bus stop, i.e., bus stops $j\,(\neq i)$ with $w_{ij} \neq 0$. The Gini coefficients computed for all bus stops in the Seoul BTN lead to the spatial distribution in Figure 3, where the Gini coefficient levels are color-coded. Overall, the spatial distributions in the years 2011 and 2013 turn out to be qualitatively the same, confirming that the Seoul BTN remained stable between these two years.

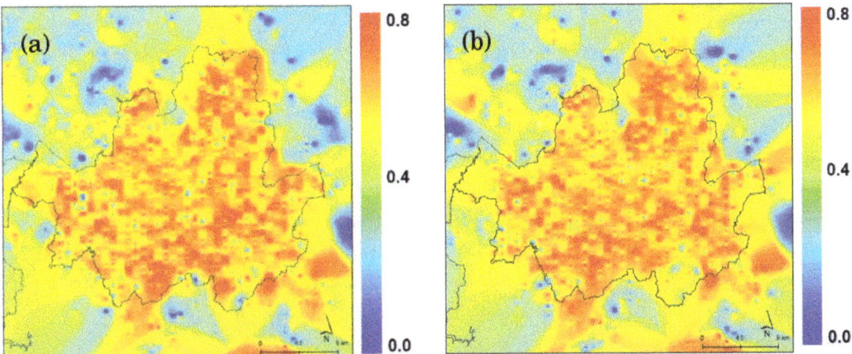


**Figure 3.** Heat map of Seoul depicting the spatial distribution of the Gini coefficient levels, specified by color, of all bus stops in the years (**a**) 2011 and (**b**) 2013. (For the color code, see the vertical bar on the right.) Accordingly, intense red regions correspond to those housing hub stops while the black line corresponds to the municipal boundary of Seoul.

Then the Gini coefficient $G_i$ of each stop $(i = 0, 1, \cdots, N)$ is sorted in the decreasing order. In plotting data, we consider the Gini coefficient $G$ at the increment of 0.01. Namely, we divide the $G$-axis into intervals of length 0.01 and take the average in each interval. The resulting distribution of the Gini coefficients for all stops is shown in Figure 4, where each data point represents the average over stops in each interval (of length 0.01) of the Gini coefficient.

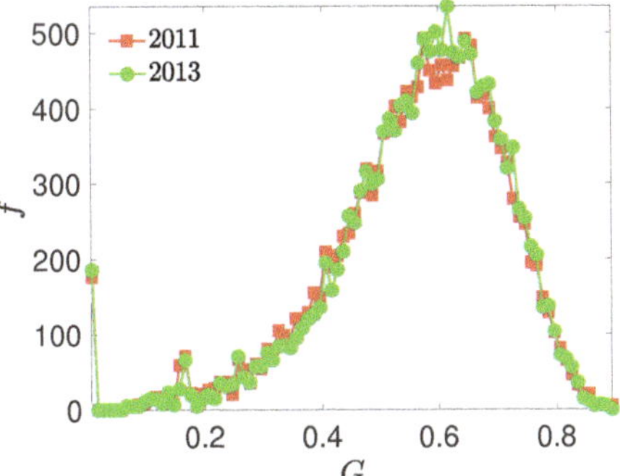


**Figure 4.** Distributions of Gini coefficients in 2011 (red) and 2013 (green). Lines are merely guides to the eye.

## 3. Results

### *3.1. Hub and Peripheral Nodes*

We make use of the Gini coefficient to decompose the BTN into two groups of nodes: hubs and peripheries. Usually, peripheral nodes are connected to only a few (two in the majority of the case) nodes, since many of such peripheral stops are served by single routes. In consequence, the weight $w_{ij}$ from a peripheral stop $i$ may not vary much with $j$ except those linked directly via the route serving $i$. This should result in a relatively small value of the Gini coefficient $G_i$. On the other hand, hubs, each served by a relatively large number of routes, accommodate many passengers with the numbers depending on the destination stops. Namely, the weight $w_{ij}$ from a hub stop $i$ is expected to vary significantly with $j$ according to whether stop $j$ is linked directly via routes serving $i$. In short, hubs and peripheries are expected to be characterized by relatively large and small values of the Gini coefficient, respectively, located in the red regions and in the blue to yellow regions in Figure 3.

To confirm this, we compute the mean weight $\bar{w}$ of each node. The mean weight $\bar{w}_i$ of stop $i$ is given by the average of weight $w_{ij}$ over the origin/destination stop $j$: $\bar{w}_i \equiv d_i^{-1} \sum_{j=1}^{d_i} w_{ij} = s_i/d_i$ with the degree $d_i$ being the number of origin and destination stops to and from stop $i$, respectively. The mean weight versus the Gini coefficient is presented in Figure 5. It is observed that the mean weight tends to increase with the Gini coefficient, which suggests that passengers using a stop tend to increase more than proportionally with the number of origins/destinations served by the stop. Since hub stops are in general expected to have larger degrees and weights (i.e., serve more destinations and passengers) in comparison with peripheral stops, it is sensible that hubs/peripheries have relatively large/small values of the Gini coefficient.

The remaining task is how to choose the boundary between hubs and peripheries. Here we notice that Figure 5 exhibits characteristic changes of the behavior at the Gini coefficient $G \approx 0.5$: Specifically, the mean weight growth with $G$ changes its slope at $G \approx 0.5$. For more information, we also consider the rank distribution of the Gini coefficient and present the result in Figure 6, where each data point represents the Gini coefficient at each rank. To probe the change in the decreasing behavior of the Gini coefficient with the rank, we perform the convexity test [39], which reveals that the convexity changes (from convex to non-convex behavior) around the rank 11,000. It is pleasing that this rank corresponds to $G \approx 0.5$ as well. This apparently suggests $G \approx 0.5$ as the boundary between hubs and peripheries. As a result, we choose specifically the rank 11,108 as the criterion separating hubs and peripheries since the bus stop ID on this rank commonly exists in both 2011 and 2013 data sets.

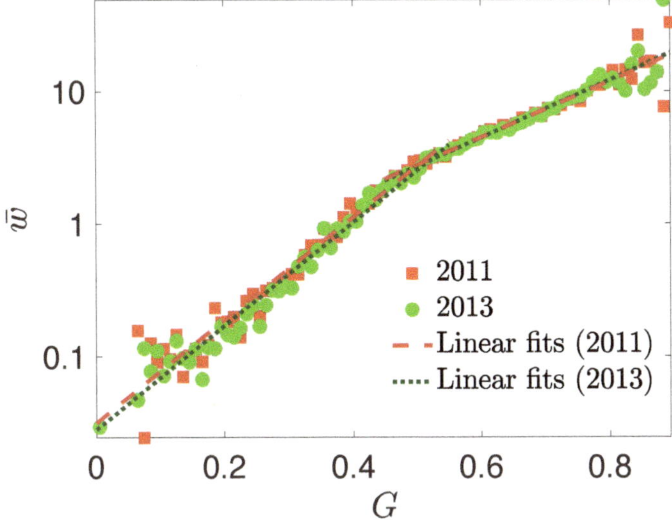


**Figure 5.** Mean weight $\bar{w}$ versus the Gini coefficient $G$ for each node. The two lines at each year, least-square fits of the data points below and above $G \approx 0.5$, meet at that point.

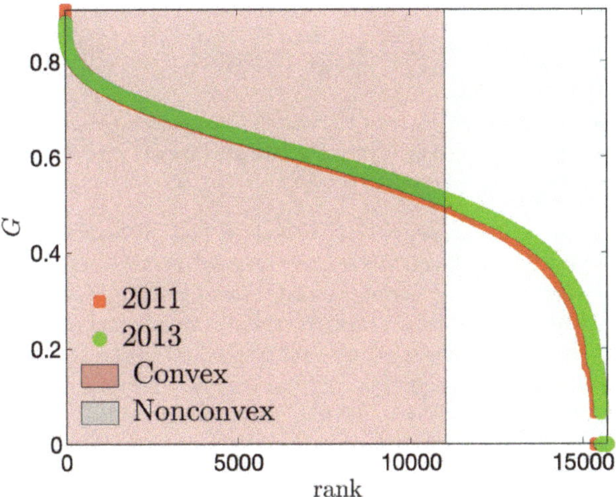


**Figure 6.** Rank distribution of the Gini coefficient. The convexity is shown to change around the rank 11,000.

### 3.2. Weight Changes Depending on the Node Type

To probe the time evolution of the Seoul BTN, we first consider the change of the weight between nodes (stops) of the two types during the two years from 2011 to 2013 and analyze the application of the Yule-type growth model to the data sets. Specifically, comparison of the data in 2011 (April 11th) and in 2013 (March 4th) shows increases both in bus stops (from 15,515 to 15,702) and in passengers (from 5,668,431 to 6,188,333). Comparing bus stop IDs in both years, we identify 8408 stops common in both years, as presented in Figure 1; these correspond to 64.1% of total stops in 2011 and 63.9% of those in 2013. In extracting common bus stops, we have considered bus stop IDs and GPS locations, and adopted the bus stop ID as the criterion. It turns out that even for the same GPS location, the stop IDs vary frequently during the two years. Reasons for such variations are not released publicly and are presumed to be policy and urban planning issues.

We now consider weights between any two among the stops, classified into the three types of pair: periphery-periphery, hub-periphery, and hub-hub pairs. Then several regression methods, including the least square regression (LSR), least absolute deviation (LAD) [40], and orthogonal regression (OR) [41–43], are applied, leading to the correlations of the three types of weight between the two years. Figure 7 exhibits the obtained weight correlations between the years 2011 and 2013. The slopes of the correlations obtained via the three regression methods are presented in Table 1.

**Table 1.** Slopes of the weight changes obtained via regression analyses of the data for the two years, 2011 and 2013.

| | LSR | LAD | OR |
|---|---|---|---|
| periphery-periphery | 0.93 | 0.91 | 0.72 |
| Hub-periphery | 0.95 | 0.97 | 0.71 |
| Hub-hub | 0.94 | 0.97 | 0.72 |

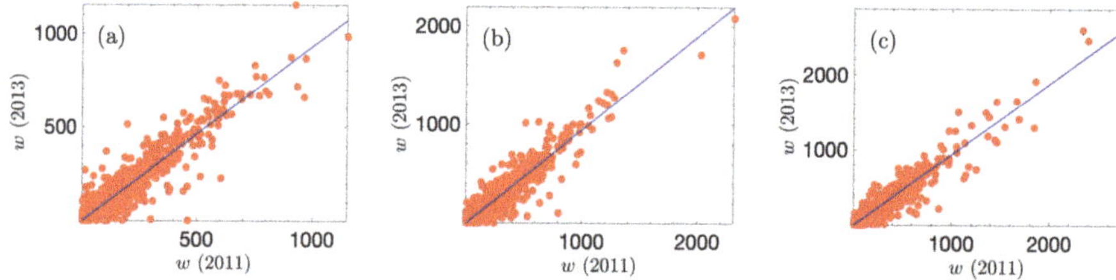


**Figure 7.** Weight correlations between the years 2011 and 2013. Weights between (**a**) periphery-periphery pairs, (**b**) hub-periphery pairs, and (**c**) hub-hub pairs. Blue solid lines plot the results of least square regression (LSR). Note the difference in the scale of the box.

The weight correlations between peripheral nodes show a bit smaller values of the slope (albeit obscure in OR), compared with the other two cases involving hub nodes. This indicates that, albeit not very conspicuous, the weights between peripheral nodes have decreased more than those involving hub nodes. One may argue that such a downward tendency in the weight between peripheral nodes is not consistent with the growth model, which usually predicts the emergence of power-law or log-normal type behavior. It is a consequence of the fact that there were far more stops in 2013 than in 2011, which reflects the construction of new stops. Despite the growth of the total strength (from 5,668,431 to 6,188,333), the average strength (per stop) of the peripheries turns out to have decreased by 30.2% while that of the hubs increased by 17.4%.

### 3.3. *Evolution of the Seoul Bus Network*

The bus stop rank sorted by the Gini coefficient provides a criterion for discriminating stops in the BTN into two distinct types: hubs and peripheries, which allows us to analyze the distributions and growth mechanisms for the two separately. Specifically, those bus stops with ranks above 11,108 are classified as hubs, while lower-ranked bus stops, having ranks below 11,108, constitute peripheral nodes.

We first examine the weight distribution of each category, which supports the validity of the Gini coefficient-based categorization. Indeed Figure 8 shows that the weight distributions fit well to either log-normal or power-law distributions. In each case, the parameters are estimated best via ordinary least squares (OLS) with the root-mean-square (RMS) deviation. For the log-normal distribution, they also agree with those via the maximum likelihood estimation (MLE). The distributions of weights involving peripheral nodes appear to follow power-law distributions: $f(x) \sim x^{-\beta}$, with the exponent $\beta = 2.17$ and 2.23 for the weights between two peripheries and between a hub and a periphery, respectively, in the year 2013. In 2011, we have $\beta = 2.18$ for the weights between two peripheries and 2.22 for those between a hub and a periphery. On the other hand, weights between hubs exhibit a log-normal distribution with the location and the scale parameters $(\mu, \sigma) = (1.74, 1.34)$ for 2011 and $(1.75, 1.37)$ for 2013.

These results indicate that while the BTN is constructed, passengers tend to make trips to or from newly constructed peripheral stops. The corresponding production of new weights gives rise to the power-law behavior of the resulting weight distributions, with the exponents $\beta$ greater than unity, according to the growth model [19]. In contrast, as to hub stops, the growth of the existing trips between them is dominant, leading to the log-normal distribution of the hub-hub weights.

Equipped with the finding from the analysis of the weight distributions, we also probe the growth mechanism for the strength distributions, shown in Figure 9. It is observed that the strength distributions of the hub category follow log-normal distributions. Notably, the deviations in the small strength regime in Figure 2 are not present. The strength distributions of peripheral stops also fit log-normal distributions in the range $s \gtrsim 100$. Table 2 summarizes the distribution parameters for the two years 2011 and 2013.

**Table 2.** Location and scale parameters, $\mu$ and $\sigma$, for the log-normal distribution, with their confidence intervals (CI) in 2011 and 2013.

| | Year | $\mu$ | CI($\mu$) | $\sigma$ | CI($\sigma$) |
|---|---|---|---|---|---|
| Hub | 2011 | 5.95 | 5.90 ∼ 5.99 | 1.51 | 1.48∼1.55 |
| | 2013 | 6.34 | 6.29∼6.39 | 1.52 | 1.48∼1.55 |
| Periphery | 2011 | 5.86 | 5.70∼5.75 | 1.46 | 1.49∼1.52 |
| | 2013 | 5.58 | 5.82∼5.89 | 1.53 | 1.44∼1.49 |

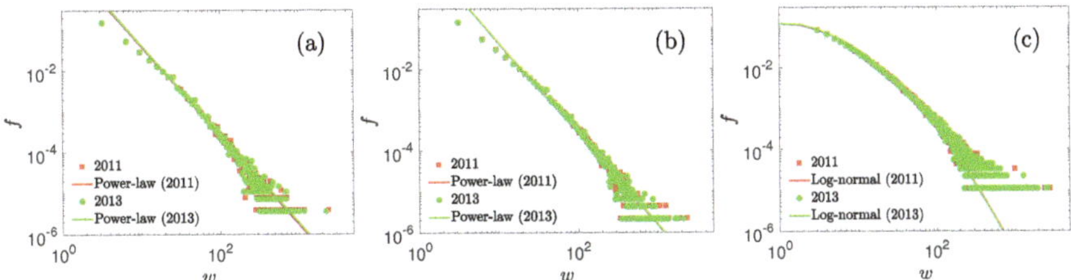


**Figure 8.** Distributions of (**a**) periphery-periphery, (**b**) hub-periphery, and (**c**) hub-hub weights. In (**a**,**b**), the red lines represent power-law distributions with exponent $\beta = 2.17$ and 2.18, respectively, for the year 2011; green lines plot power-law distributions for 2013, with exponent $\beta = 2.23$ and 2.22, respectively. In (**c**) the red and green lines depict the log-normal distributions with $(\mu, \sigma) = (1.74, 1.34)$ for 2011 and $(1.75, 1.37)$ for 2013, respectively. In all cases, the two lines overlap with each other too closely to distinguish.

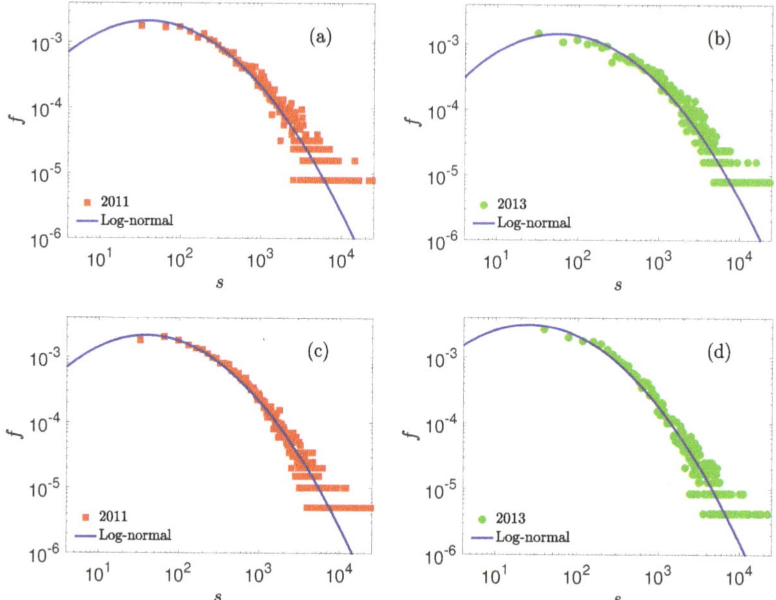


**Figure 9.** Strength distributions on the log-log scale. Hub nodes in (**a**) 2011 and (**b**) 2013; peripheral nodes in (**c**) 2011 and (**d**) 2013. Solid lines plot the log-normal distributions with parameters given in Table 2.

In the perspectives of the controlled growth model, the emergence of log-normal distributions may imply the presence of a certain evolution mechanism such as growth and production in the BTN. In the case of hub nodes, time evolution is governed by the growth of each node; on the other hand, production (birth) of new nodes should be taken into account to understand the time evolution of peripheral nodes.

To probe the time evolution of the Seoul BTN, we consider the bus stops having the same ID in Figure 1. It is known that the Yule-type growth (possibly with self-size production) results in the log-normal distribution [26,29]:

$$f(s,t) = \frac{1}{\sqrt{2\pi}\sigma_t s} \exp\left[-\frac{(\ln s - \mu_t - \mu_0)^2}{2\sigma_t^2}\right], \tag{3}$$

where the location parameter $\mu$ and the scale parameter $\sigma$ evolve in time according to

$$\begin{aligned} \mu_t &= \lambda t \ln(1+b) \\ \sigma_t &= \sqrt{\lambda t}\ln(1+b) \end{aligned} \tag{4}$$

with the (mean) growth rate $\lambda$, growth factor $b$, and initial location $\mu_0$.

We now examine the Seoul BTN in view of Equation (3), to check the applicability of the Yule-type growth model [26]. Analyzing the strength data for the hubs in the two years 2011 and 2013, we find the growth rate $\lambda = 0.496$ and growth factor $b = 0.129$. Since we focus on bus stops common in both years, it is natural that the growth probability for two years, given by $\lambda\Delta t$ with $\Delta t = 2$ (years), takes the value close to unity. Plugging these values into Equation (4), we obtain the changes in the location parameter and the scale parameter during two years: $\Delta\mu = \lambda\Delta t \ln(1+b) = 0.120$ and $\Delta\sigma^2 = \lambda\Delta t[\ln(1+b)]^2 = 0.0146$. These theoretical results are to be compared with the results from the passenger data, $\Delta\mu = 0.39$ and $\Delta\sigma^2 = 0.0097$. Unlike the scale parameter, the location parameter shows a substantial discrepancy between the theoretical result and passenger data. In consideration of the limited data with the quite small growth factor, however, the two results, one from theoretical analysis and the other from passenger data, are presumed reasonably consistent with each other.

On the other hand, the growth of peripheral nodes is not captured properly by the controlled growth model. In particular, the peripheries display rather shrinking behavior than growth, as the passenger data have the mean $\mu = 5.58$ in 2013, decreased from $\mu = 5.86$ in 2011. Such different behaviors between hubs and peripheries can be attributed to the difference in bus stops of common IDs over the two years. Among those common stops, hub nodes are fully consistent over the years 2011 and 2013. Accordingly, each hub is subject to growth, albeit without newly-constructed bus stops, and the set of hubs is described by the controlled growth model, giving rise to the log-normal distribution. In contrast, only about half of the peripheral nodes persisted and are included in the set of common stops. To be precise, 48.71% of the peripheries present in 2011 were closed during the two years from 2011 to 2013 while newly constructed nodes during the two years occupied 48.04% of the total peripheries in 2013. We also attempt to match those peripheral stops of changed IDs with the help of their GPS locations, which turns out to give about 50% of stops persistent as well; in the case of hubs, this percentage of unchanged stops reaches 65% of the total. This analysis manifests that bus stops having small values of the Gini coefficient change frequently due to the low usage.

This also provides, as a by-product, a good reason to name this set of nodes *peripheral*, which are vulnerable to frequent changes at the policy level. Planning policies could include various government-creating or artificial transitions to remove less-visited stops. As Seoul is a very mature metropolitan city, its downtown area is already sufficiently saturated [19]. In consequence, closing of less used stops or construction of new stops is naturally concentrated in peripheral parts of the city. It should also be noted that the peripheral parts are not necessarily confined within suburban areas; stops in the area with less-developed transportation are possibly categorized as peripheral nodes.

## 4. Discussion

The BTN has unique characteristics, which may not be easily captured as in other transportation systems such as subways or airlines. Serving as a dominant transportation mode for short- to mid-distance trips, the BTN is important for policymakers to deal with the complexity of urban mobility and make more sustainable city planning, which should impact on both passengers and policymakers. To apply sustainability analysis with better accuracy, one needs to understand the BTN concerning not only distribution but also growth.Unlike other transportation networks, the passenger flow in the BTN may not be described plausibly by a single distribution, which reflects that there are two distinct types of bus stops and the trips thus vary according to the type involved.

In this study, we have considered how to characterize the BTN with regard to its evolution in time; this is expected to provide a fundamental approach to a better understanding of its structure, which should help sustainability assessment. Possible candidates for characterization include locations of bus stops, the strength of each stop, weight between a pair of stops, passenger concentration, and so on. We have here proposed to use the Gini coefficient, which measures disparity among weights of trips between each pair of nodes. Making use of the Gini coefficient for each bus stop, we have classified the stops into two distinct types: hub and peripheral nodes. Separating hubs having higher ranks and peripheries with lower ranks, we have observed log-normal distributions for both sets of nodes.

Specifically, comparison of the data in the years 2011 and 2013 has shown strengths of the hubs to grow, which is well explained by the Yule-type growth model. Analyzing the passenger data in the Seoul BTN, we have obtained parameters for the model. It has thus been concluded that the controlled growth model well captures the evolution of the set of hub nodes: It provides a good description, not only qualitatively but also quantitatively, of such features as the log-normal distribution, evolving through the mean and the deviation in time.

On the other hand, peripheral nodes are vulnerable to frequent changes due to low usage, which makes it inappropriate to apply the growth model. In Seoul, where downtown areas are already substantially saturated, changes in the BTN occur more often in peripheral nodes. To summarize, we point out that the Gini-based approach captures the key characteristics of the BTN among other transportation networks. It would be desirable to develop an algorithm to match closed bus stops with newly-constructed ones. Here, appropriate criteria may include not only the GPS locations of stops but also some policy-level information. This is left for future study.

**Author Contributions:** Conceptualization, K.L. and M.Y.C.; data curation, C.K., S.G., M.S.C., and K.L.; formal analysis, C.K., S.G., and M.S.C.; funding acquisition, K.L. and M.Y.C.; investigation, K.L. and M.Y.C.; methodology, S.G.; project administration, M.Y.C.; resources, K.L.; software, C.K., S.G., and M.S.C.; supervision, M.Y.C.; validation, C.K. and K.L.; writing—original draft, S.G. and K.L.; writing— review and editing, C.K. and M.Y.C. All authors have read and agreed to the published version of the manuscript.

**Funding:** This work was supported by the Sungshin Women's University Research Grant of 2018 (KL), by the National Research Foundation of Korea through the Basic Science Research Program (Grant No. 2019R1F1A1046285) (MYC), and by Korea Institute of Science and Technology through the Economic and Social Science Research Initiative (ESRI) 2E30443 and Pride-up 2V08460 (CK).

**Conflicts of Interest:** The authors declare that they have no competing interests.

## References

1. Watts, D.; Strogatz, S. Collective of small-world networks. *Nature* **1998**, *393*, 440–442. [CrossRef]
2. Barabási, A.L.; Albert, R. Emergence of scaling in random networks. *Science* **1999**, *286*, 509–512. [CrossRef] [PubMed]
3. Acemoglu, D.; Carvalho, V.M.; Ozdaglar, A.; Tahbaz-Salehi, A. The network origins of aggregate fluctuations. *Econometrica* **2012**, *80*, 1977–2016. [CrossRef]
4. Jackson, M.O. *Social and Economic Networks*; Princeton University Press: Princeton, NJ, USA, 2008.
5. Pardee, F. United States. Northeast Corridor Transportation Project. In *Measurement and Evaluation of Transportation System Effectiveness*; Rand Corporation Memorandum RM-5869-DOT; Rand Corporation: Santa Monica, CA, USA, 1969.
6. Altshuler, A.; Womack, J.; Pucher, J. *The Urban Transportation System: Politics and Policy Innovation*; MIT Press Series in Transportation Studies; MIT Press: Cambridge, MA, USA, 1979.
7. Dubois, D.; Bel, G.; Llibre, M. A set of methods in transportation network synthesis and analysis. *J. Oper. Res. Soc.* **1979**, *30*, 797–808. [CrossRef]
8. Zegras, C. Sustainable transport indicators and assessment methodologies–Biannual Conference and Exhibit of the Clean Air Initiative for Latin American Cities. In Proceedings of the Sustainable Transport: Linkages to Mitigate Climate Change and Improve Air Quality, São Paulo, Brazil, 25–27 July 2006.
9. Hu, Y.; Zhu, D. Empirical analysis of the worldwide maritime transportation network. *Physica A* **2009**, *388*, 2061–2071. [CrossRef]
10. Canas, G.; Rosasco, L. Learning probability measures with respect to optimal transport metrics. In *Advances in Neural Information Processing Systems 25*; Pereira, F., Burges, C.J.C., Bottou, L., Weinberger, K.Q., Eds.; Curran Associates, Inc.: New York, NY, USA, 2012; pp. 2492–2500.
11. Farahani, R.Z.; Miandoabchi, E.; Szeto, W.; Rashidi, H. A review of urban transportation network design problems. *Eur. J. Oper. Res.* **2013**, *229*, 281–302. [CrossRef]
12. Zietsman, J.; Rilett, L.; Kim, S. *Sustainable Transportation Performance Measures for Developing Communities*; Technical Report; Texas Transportation Institute, Texas A&M University System: San Angelo, TX, USA, 2003.
13. Kim, H. *Developing Guidelines for Evaluating and Selecting Urban Rail Transit Systems*; The Korea Transport Institute: Seoul, Korea, 2008.
14. Jeon, C.M.; Amekudzi, A.A.; Guensler, R.L. Sustainability assessment at the transportation planning level: Performance measures and indexes. *Transp. Policy* **2013**, *25*, 10–21. [CrossRef]
15. Liu, X.; Derudder, B.; Wu, K. Measuring polycentric urban development in China: An intercity transportation network perspective. *Reg. Stud.* **2016**, *50*, 1302–1315. [CrossRef]
16. Lee, K.; Jung, W.S.; Park, J.S.; Choi, M.Y. Statistical analysis of the Metropolitan Seoul subway system: Network structure and passenger flows. *Physica A* **2008**, *387*, 6231. [CrossRef]

17. Lee, K.; Park, J.S.; Choi, H.; Jung, W.S.; Choi, M.Y. Sleepless in Seoul: The ant and the metrohopper. *J. Korean Phys. Soc.* **2010**, *57*, 823.
18. Lee, K.; Goh, S.; Park, J.S.; Jung, W.S.; Choi, M.Y. Master equation approach to the intra-urban passenger flow and application to the metropolitan Seoul subway system. *J. Phys. A* **2011**, *44*, 115007. [CrossRef]
19. Goh, S.; Lee, K.; Park, J.S.; Choi, M.Y. Modification of the gravity model and application to the metropolitan Seoul subway system. *Phys. Rev. E* **2012**, *86*, 026102. [CrossRef] [PubMed]
20. Yang, X.H.; Chen, G. Bus transport network model with ideal n-depth clique network topology. *J. Phys. A* **2011**, *390*, 4660–4672.
21. Goh, S.; Lee, K.; Choi, M.Y.; Fortin, J.Y. Emergence of criticality in the transportation passenger flow: Scaling and renormalization in the Seoul bus system. *PLoS ONE* **2014**, *9*, e89980. [CrossRef] [PubMed]
22. Lee, K.; Park, J.S.; Goh, S.; Choi, M.Y. Accessibility measurement in transportation networks and application to the Seoul bus system. *Geograph. Anal.* **2019**, *51*, 339. [CrossRef]
23. Stenneth, L.; Wolfson, O.; Yu, P.S.; Xu, B. Transportation Mode Detection Using Mobile Phones and GIS Information. In Proceedings of the 19th ACM SIGSPATIAL International Conference on Advances in Geographic Information Systems, Chicago, IL, USA, 1–4 November 2011; Association for Computing Machinery: New York, NY, USA, 2011; pp. 54–63.
24. Badia, H. Comparison of bus network structures in face of urban dispersion for a ring-radial city. *Netw. Spat. Econ.* **2020**, *20*, 233–271. [CrossRef]
25. Choi, M.Y.; Choi, H.; Fortin, J.Y.; Choi, J. How skew distributions emerge in evolving systems. *EPL* **2009**, *85*, 30006. [CrossRef]
26. Goh, S.; Kwon, H.W.; Choi, M.Y.; Fortin, J.Y. Emergence of skew distributions in controlled growth processes. *Phys. Rev. E* **2010**, *82*, 061115. [CrossRef]
27. Choi, J.; Choi, M.Y.; Yoon, B.G. Interaction effects on the size distribution in a growth model. *J. Korean Phys. Soc.* **2018**, *72*, 327. [CrossRef]
28. Kwon, H.W.; Choi, M.Y. How cells grow and divide: Mathematical analysis confirms demand for the cell cycle. *J. Phys. A* **2012**, *45*, 135101. [CrossRef]
29. Kim, C.; Kim, D.S.; Ahn, K.; Choi, M.Y. Dynamics of analyst forecasts and emergence of complexity: Role of information disparity. *PLoS ONE* **2017**, *12*, e0177071. [CrossRef] [PubMed]
30. Gini, C. *Variabilitá e Mutabilitá*; Libreria Eredi Virgilio Veschi: Rome, Italy, 1912.
31. Gini, C. Measurement of inequality of incomes. *Econ. J.* **1921**, *31*, 124–126. [CrossRef]
32. Lambert, P.J.; Aronson, J.R. Inequality decomposition analysis and the Gini coefficient revisited. *Econ. J.* **1993**, *103*, 1221–1227. [CrossRef]
33. Dorfman, R. A formula for the Gini coefficient. *Rev. Econ. Stat.* **1979**, *61*, 146–149. [CrossRef]
34. Giger, E.; Pinzger, M.; Gall, H. Using the Gini coefficient for bug prediction in eclipse. In Proceedings of the 12th International Workshop on Principles of Software Evolution and the 7th Annual ERCIM Workshop on Software Evolution, Szeged, Hungary, 5–6 September 2011; IWPSE-EVOL '11, ACM: New York, NY, USA, 2011; pp. 51–55. [CrossRef]
35. Mussard, S.; Terraza, M.; Seyte, F. Decomposition of Gini and the generalized entropy inequality measures. *Econ. Bull.* **2003**, *4*, 1–6.
36. Lisker, T. Is the Gini coefficient a stable measure of galaxy structure? *Astrophys. J. Suppl. Ser.* **2008**, *179*, 319–325. [CrossRef]
37. Sundhari, S. A knowledge discovery using decision tree by Gini coefficient. In Proceedings of the 2011 International Conference on Business, Engineering and Industrial Applications, Kuala Lumpur, Malaysia, 5–7 June 2011; pp. 232–235.
38. Sanasam, R.; Murthy, H.; Gonsalves, T. Feature selection for text classification based on Gini coefficient of inequality. In Proceedings of the Fourth International Workshop on Feature Selection in Data Mining, Hyderabad, India, 21 June 2010; Liu, H., Motoda, H., Setiono, R., Zhao, Z., Eds.; PMLR: Hyderabad, India, 2010; Volume 10, Proceedings of Machine Learning Research, pp. 76–85.
39. Boyd, S.; Vandenbergh, L. *Convex Optimization*; Cambridge University Press: New York, NY, USA, 2004; pp. 324–342.
40. Pollard, D. Asymptotics for least absolute deviation regression estimators. *Econom. Theory* **1991**, *7*, 186–199. [CrossRef]

41. Carroll, R.J.; Ruppert, D. The use and misuse of orthogonal regression in linear errors-in-variables models. *Am. Stat.* **1996**, *50*, 1–6.
42. Maronna, R. Principal components and orthogonal regression based on robust scales. *Technometrics* **2005**, *47*, 264–273. [CrossRef]
43. Leng, L.; Zhang, T.; Kleinman, L.; Zhu, W. Ordinary least square regression, orthogonal regression, geometric mean regression and their applications in aerosol science. *J. Phys. Conf. Ser.* **2007**, *78*, 012084. [CrossRef]

*Article*

# Investigate Tourist Behavior through Mobile Signal: Tourist Flow Pattern Exploration in Tibet

**Lina Zhong [1], Sunny Sun [2,*], Rob Law [3] and Liyu Yang [4]**

1. Institute for Big Data Research in Tourism, School of Tourism Sciences, Beijing International Studies University, Beijing 100020, China; zhonglina@bisu.edu.cn
2. College of Asia Pacific Studies, Ritsumeikan Asia Pacific University, Beppu, Oita 874–8577, Japan
3. School of Hotel & Tourism Management, The Hong Kong Polytechnic University, Hong Kong, China; rob.law@polyu.edu.hk
4. School of Tourism Sciences, Beijing International Studies University, Beijing 100020, China; liyu_yang@163.com

* Correspondence: sunnysun@apu.ac.jp

Received: 23 September 2020; Accepted: 30 October 2020; Published: 3 November 2020

**Abstract:** Identifying the tourist flow of a destination can promote the development of travel-related products and effective destination marketing. Nevertheless, tourist inflows and outflows have only received limited attention from previous studies. Hence, this study visualizes the tourist flow of Tibet through social network analysis to bridge the aforementioned gap. Findings show that the Lhasa prefecture is the transportation hub of Tibet. Tourist flow in the eastern part of Tibet is generally stronger than that in the western part. Moreover, the tourist flow pattern identified mainly includes "(diverse or balanced) diffusion from the main center", "clustering to the main center", and "diffusion from a clustered circle".

**Keywords:** pattern; social network analysis; tourist flow; visualization; Tibet

---

## 1. Introduction

Tourist flow refers to the spatial distribution of tourists, reflecting the travel patterns of tourists in a certain region [1]. Understanding the spatial distribution of tourist flows and the movement patterns of tourists can provide practical implications to tourism practitioners in terms of resource allocation, infrastructure construction, and effective tourism planning for a destination [2–5]. Tourist flow can also assist in the management of tourism's environmental and cultural impacts [4]. Cox found three movement patterns of humans; namely, distance-, direction-, and connection-based movement patterns [6]. The distance-biased movement pattern denotes the distance-related intensity of movement. The direction-based movement pattern reflects the direct movement of tourists. Last, the connection-biased movement pattern pays attention to the connection points during movement, reflecting the important role of connection in determining the characteristics of the movement. Oppermann revealed that "trip itineraries" can be used to reflect travel patterns and tourist flows. In addition, Oppermann proposed two categorizations of tourist movements; namely, tourist movement among different locations and tourists' stays in different locations [7]. Zhong, Zhang, and Li pointed out the disadvantages of conventional directional bias, such as the inability to identify the emitted and the attracted tourist flows, and proposed a new concept of "functional tourism region" [8]. Previous investigations of tourist flows are mostly based on scale. For example, Liu, Zhang, Zhang, and Chen categorized 31 provincial destinations in China into national and regional tourist centers and common and marginal destinations [9]. Jin, Xu, Huang, and Cao investigated tourist flows in different attractions in Nanjing, the inner city of China [10]. In summary, the identification of movement patterns of tourists seems to shift from complex to simple, and the investigation of the scale of the spatial distribution of tourists ranges from the country to the city level [2,11].

Familiarity with tourist flow can guide tourism practitioners to achieve effective and harmonious coordination in different areas of a destination and ultimately promote sustainable tourism development [9]. Tourist flows can also assist tourism practitioners in identifying the potential of a particular destination and promoting balanced tourism development. With the enhanced accessibility brought by transportation, tourist travel is no longer restricted to the most commonly recommended or popular tourist attractions. Tibet, a remote and autonomous region of China, has been developing rapidly because of enhanced accessibility and China's national development strategy of turning the plateau into a tourism destination [12]. Tourist arrivals and tourism revenues in Tibet have increased by more than 20% since the opening of the Qinghai–Tibet railway in 2006 [13,14]. In addition, the rapid development of the Internet increases Tibet's opportunities to connect with the outside world [5]. Hence, Tibet may further develop its tourism industry. As a large destination with different prefectures, the present study selected Tibet as a case to generate tourist flow patterns from a regional perspective, together with the detailed tourist flow patterns in each of its prefectures. Hence, Tibet is selected as a case to help tourism practitioners come up with corresponding strategic plans for a balanced and sustainable tourism development [12].

Although previous studies examined tourist movement patterns or categorized tourism destinations from a country or city level [9,10], the implications of the country- or city-level tourist flow is either very general or specific. The detailed implications of tourist flow to a regional-level perspective are also limited, to a certain extent. In addition, although previous studies explored tourists' movement patterns or itineraries, tourist inflows and outflows received limited attention. Knowledge of tourist inflows and outflows can greatly assist destinations in balancing regional tourism development and developing their tourism effectively. Hence, the present study uses the theory of social network analysis (SNA) to identify the tourist flows in each of Tibet's prefecture. Specifically, the present study aims to visualize tourist flow in Tibet and examine the differences among tourist inflows, outflows, and total tourist flow. Moreover, this study aims to summarize the patterns of tourism flows and provide practical implications for effective future tourism planning and development.

## 2. Literature Review

### 2.1. Tourist Flow or Movement

Investigations on tourist flow have begun since the late 1980s to 1990s, mainly focusing on economic impacts. For example, Japan received limited tourism inflows despite its massive tourism outflows. Buckley, Mirza and Witt indicated that further attention should be paid to the structural adjustment to attract foreign travelers to Japan and balance the payment of deficit items [15]. Kulendran adopted a cointegration analysis to estimate quarterly tourist flows to Australia from countries such as the US and Japan, to investigate the economic impacts of tourist flows. Moreover, Kulendran considered the factors affecting travel, such as income and airfare, and found that the estimated long-run elasticity of the relative price variable for the UK and Japan is close to unity but is greater than one for the US and New Zealand [16]. Jansen-Verbeke and Spee examined the interregional and intraregional tourist flows within Europe and found that competition does not exist among countries but among regions [11].

Coshall adopted spectral analysis to detect cycles within and between time-series datasets and found that the leading cycles of dependencies rely on tourist flows and exchange rates [17]. Dimoska and Petrevska considered the net flows of tourism services by analyzing the balance of payments in Macedonia and revealed possible ways to increase tourism inflows, such as attracting foreign tourists [18]. In addition, Patuelli, Mussoni, and Candela used an econometric model and found that regions with world heritage sites in Italy can enhance tourist inflows, considering all else being equal [19]. Degen examined nine main stations of the Beijing–Shanghai High-Speed Railway through

SNA and identified the destination choice of tourists, spatial distribution, and travel time [1]. Then, Degen compared the time-space distribution of tourist flow and found that tourism origins—Beijing, Shanghai, and Nanjing—are strengthened as tourism centers, and tourism flow resulted in the "Matthew effect" [1]. Here, the "Matthew effect" reflects the cumulative advantage of main cities as tourism centers.

Lew and McKercher [4] identified the factors that could influence the movement patterns of tourists and found that such patterns are influenced by four types of territorial and three linear path models. Specifically, the four types of tourist movement behaviors in a destination are "no movement," "convenience-based movement," "concentric exploration," and "unrestricted destination wide movement" [4]. They identified three movement pattern variations of tourists; namely, "point-to-point," "circular," and "complex patterns" [4]. Zhong et al. studied the overnight travel patterns of Chinese tourists and categorized tourists flow into pointing of tourist flow and inertial state of pointing [8]. Three basic types of pointing were identified, namely, "city," "seaside," and "sun-lust pointing." Based on the extended contemporary urban transportation model, Zhong et al. also identified the movement patterns of tourists to examine and enhance the attractiveness of a destination [5]. Data about tourist movement or flow in a tourist destination are not enough to provide detailed implications regarding tourism development in a destination. Hence, the flow patterns of tourists in a certain region must be explored further to provide valuable implications for sustainable tourism development.

### 2.2. Theory of SNA and Its Extension in Tourism

Diverse analytical methods are used to study tourists' movement patterns, including co-integration analysis, spectral analysis, and SNA [1,16,17]. The present study selects SNA because it can evidently reflect the nodes during the changes or movements of tourists, which provides an overall picture of the movement or flow patterns rather than the characteristics of individuals only [20]. The SNA concept appeared before the 1930s, and researchers gradually built the concept of social structure and recognized its importance as a social "fabric" and "web" [21]. Wasserman and Faust [20] noted that SNA examines the relationships among people, organizations, or other related information and is normally reflected as links or nodes to form a network. Otte and Rousseau also confirmed that SNA can be used to investigate social structures [22].

Shih investigated the travel routes of drive tourists from 16 destinations in Nantou, Taiwan, and provided the structural patterns of connected systems by applying SNA to tourism [23]. The findings show that necessary facilities should be provided to different locations in Nantou, Taiwan [23]. Leung et al. used SNA to examine 500 online trip diaries and found tourist movement patterns during the Olympics in Beijing in 2008. The findings revealed that international tourists are mostly interested in well-known traditional attractions, and their activities are within the central areas of Beijing [24]. Raisi, Baggio, Barratt-Pugh, and Willson analyzed 1515 tourism websites and visualized the network structure using SNA. They found that the SNA structure tends not to be hierarchical, and the communities tend to be formed on the basis of the geographical locations [25]. David-Negre, Hernández, and Moreno-Gil examined the spending pattern of tourists through SNA [26]. Jin et al. found and summarized three movement patterns of tourists; namely, "diffusion from a single center," "clustering to a single center," and "balancing between multiple centers" [10].

The literature has been constantly paying attention to the economic impacts of tourist flow since the 1990s. Most previous studies only investigated tourist flow at the country level, and detailed implications to a region or a destination are lacking to a certain extent. If the scale of the destination is small with few attractions, then movement patterns of tourists/tourist flow can be easily identified. By contrast, valuable practical implications can be provided for large-scale destinations by identifying the movement patterns of tourists or tourist flow, which is more difficult in this context than that of small-scale destination [27]. Although Lew and McKercher comprehensively indicated the movement patterns of tourists, the current study focused on patterns between the locations of tourists' accommodations and attractions [4]. Lew and McKercher listed different types of tourist movements and pointed out the factors, such as demand and supply and transportation networks [4]. However, empirical studies lack data on tourist flows, which can help visualize how different regions interact and compete with one another [19]. The present study visualizes and examines the tourist flow in different prefectures in Tibet to provide not only an overall trend of tourist flow but also the flow patterns of tourists in each of the prefectures. The objective is to promote sustainable tourism development and maximize tourism revenue.

## 3. Methodology

Most previous studies adopted questionnaire surveys or secondary data to explore tourist flows or patterns in a destination [19,26]. For instance, Zeng identified the characteristics of tourist flow of Chinese tourists visiting Japan by retrieving 430 travel itineraries from travel agencies in China and 458 itineraries of independent tourists from their trip diaries [28]. Nevertheless, the aforementioned traditional data collection methods, such as questionnaire surveys or secondary data, lack efficiency and completeness. Toha and Ismail discussed the applicability of various tracking technologies, such as global positioning system or land-based tracking technologies, to track the movement of tourists in historical cities of Melaka. However, an empirical investigation is still absent [29].

Hence, considering the shortages of traditional data collection, such as questionnaire surveys, the present study employed the concurrent data collection method through the mobile signal to obtain rich data to gain comprehensive information of tourist flow. Recently, along with the wide adoption of smartphones by tourists during travel, large volumes of user-generated data have become available to generate rich data [5]. The present study tracked the movements of tourists in Tibet by retrieving data via mobile phone signals of tourists. That is to say, mobile phone signal information from China Unicom telecommunication service was retrieved. In other words, for tourists who used the telecommunication service from China Unicom, their movements in Tibet were tracked. Specifically, once tourists entered Tibet, their mobile phone signals were identified and tracked through their points of entry. Similarly, once tourists left Tibet, points of exit were recorded. In the meantime, data were encrypted to protect the privacy of tourists, and the encryption assures that the personal information of tourists was kept confidential as no party can read or obtain this information. As a result, only their movements in Tibet were tracked and recorded.

In summary, in October and November 2018, tourist inflows and outflows of different prefectures (including domestic and international tourists) in Tibet were retrieved. October was selected because China's National Day is in this month, which has a long holiday and is the peak season of tourism in Tibet. On the contrary, compared with October, November is the off-peak season for tourism in Tibet. Thus, the present study selected the representative data in October and November as an example to investigate the tourist flow pattern in peak and off-season in Tibet. The tourist flows of six regions (i.e., from west to east Tibet); namely, Ngari, Nagqu, Xigaze, Lhasa, Shannan, and Nyingchi prefectures, were retrieved based on the administrative regions in Tibet. Qamdo prefecture is not included because it received lower tourist flows compared with that in other prefectures of Tibet.

## 4. Findings and Discussion

Figure 1 depicts the directions of tourist inflows, outflows, and the total tourist flow, including the intensity of tourist flows, using software UCIENT and NETDRAW. The light blue circles represent 59 districts among six prefectures in Tibet, whereas the three dark blue circles signify tourist inflows, outflows, and the total tourist flow. The degree of the thickness of the lines represents the intensity of the tourist flow. The findings show that Lhasa prefecture (i.e., Tibet) is Tibet's transportation hub. In addition, Nagqu (i.e., Nagqu) and Xigaze (i.e., Samzhubze) prefectures are considered secondary transportation hubs. Furthermore, Nyingchi (i.e., Nyngchi) and Ngari (i.e., Gar) prefectures are supportive transportation hubs.

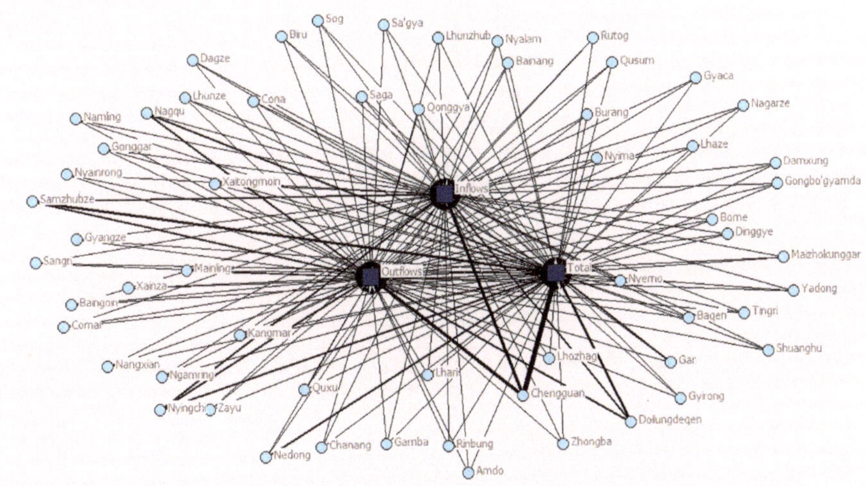


**Figure 1.** Direction and intensity of tourist flow.

Figure 2 further indicates tourist flow patterns and directions in different Tibetan prefectures. The thickness of the lines represents the amount of tourist flows in different prefectures. The degree of the thickness of the lines represents the intensity of the tourist flows. Overall, tourist flow patterns are diversified among different prefectures in Tibet. Specifically, compared with other prefectures, the Lhasa prefecture received the most tourist flows, followed by Nagqu, Xigaze, Nyingchi, Shannan, and Ngari prefectures. The following sections will provide detailed information about the number of tourist inflows or outflows, present the number of total tourist flows in each prefecture in Tibet, compare the differences of tourist inflows and outflows, and identify the flow patterns of tourists.

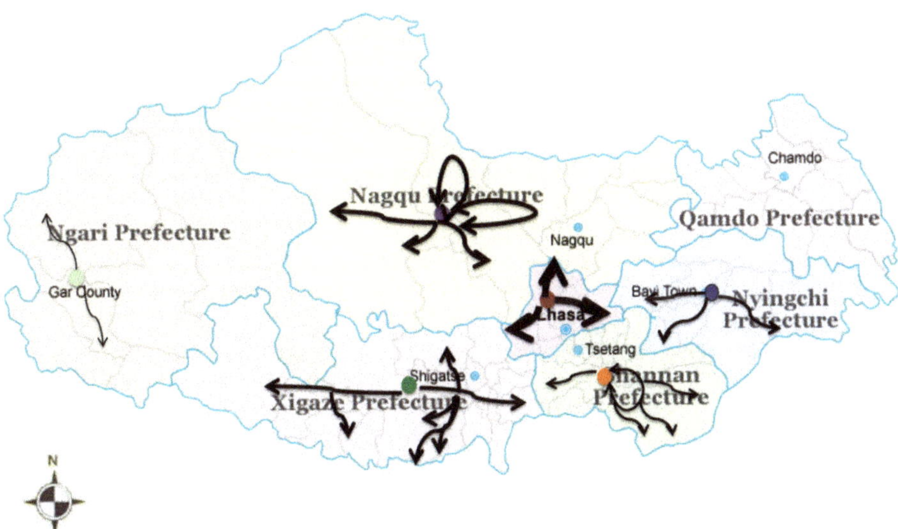


**Figure 2.** Tourist flows in different prefectures in Tibet. Note: Map of Tibet was retrieved from http://www.chinatourmap.com/tibet/tibet-political-map.html.

Table 1 shows the number of tourist inflows and outflows and the total number of tourist flows in the Ngari prefecture in Tibet. Compared with other prefectures in Tibet, the Ngari prefecture has relatively fewer tourist flows. The Gar district is a central place that connects the northern and southern parts. In addition, no significant differences are found among tourist inflows, outflows, and the total number of tourist flows in the Ngari prefecture in October and November of 2018. Located in the western part of Tibet, the Ngari prefecture generally receives fewer tourists than its eastern parts. The tourist flow pattern in the Ngari prefecture is very simple which is indicated by "a three-point line." Hence, tourism practitioners should consider the selling point of the Ngari prefecture, market its attractions, further encourage the exploration of tourists, and increase its tourist flow in different parts of the Ngari prefecture gradually.

**Table 1.** Tourist flow in Ngari prefecture.

| Ngari | Inflows (October) | Inflows (November) | Outflows (October) | Outflows (November) | Total (October) | Total (November) |
|---|---|---|---|---|---|---|
| Gar | 210,981 | 139,296 | 539,797 | 716,301 | 750,778 | 855,597 |
| Burang | 14,087 | 5015 | 112,997 | 155,945 | 127,084 | 160,960 |
| Rutog | 1994 | 2337 | 66,248 | 98,390 | 68,242 | 100,727 |
| Mean | 75,687.08 | 48,882.70 | 239,680.79 | 323,545.34 | 315,367.87 | 372,428.04 |
| Pair mean | 26,804.38 | | −83,864.55 | | −57,060.17 | |
| t | 1.186 | | −1.806 | | −2.389 | |
| df | 2 | | 7 | | 7 | |
| Sig. (2-tailed) | 0.357 | | 0.213 | | 0.139 | |

df = Degrees of freedom.

Table 2 lists the number of tourist inflows and outflows and the total number of tourist flows of the Nagqu prefecture in Tibet. Nagqu generally plays a central role in connecting different parts of the Nagqu prefecture. Specifically, the Shuanghu district in the western part of the Nagqu prefecture receives fewer tourist flows than other areas in the Nagqu prefecture. In addition, tourist flows are concentrated in the central and eastern parts of Nagqu and are scattered in western parts, such as the Xainze district. Through paired samples t-test, significant differences ($p = 0.041$) are found between October 2018 and November 2018 regarding tourist inflows of the Nagqu prefecture. They are significant at the 95% confidence interval. Specifically, the number of tourist inflows of the Nagqu prefecture in October 2018 is higher than those in November 2018. This result indicates that, for the Nagqu prefecture, compared with November, tourists prefer to visit Nagqu in October. In other words, efforts can be made by the Tibet tourist bureau to attract more tourists to visit Tibet in November by creating special themes, as an example.

Significant differences are found regarding tourist inflows of the Nagqu prefecture, whereas no significant difference is found between tourist outflows and the total number of tourist flow. The tourist flow of the Nagqu prefecture is characterized and reflected by the primary flows around the city center, along with the secondary flows between the core and the minor nodes (i.e., Nagqu–Amdo–Nyainrong; Nyainrong–Baqen–Nagqu). The tourist flow in Nagqu prefecture is generally indicated by the structure of "diffusion from the main center" and "clustering to the main center." Patuelli et al. [19] indicated that an increase in world heritage sites cannot only lead to a 4% increase in tourist inflows but also helps a certain region gain competitive advantages over other regions or districts. Hence, taking advantage of the Nagqu center and highlighting the appeal of attractions in nearby districts to extend the primary flows and increase the secondary flows can be considered. Lew and McKercher suggested that tourists can venture further as they become familiar with a region, thereby helping a destination increase secondary flows [4].

**Table 2.** Number of tourist flow in Nagqu prefecture.

| Nagqu | Inflows (October) | Inflow (November) | Outflows (October) | Outflows (November) | Total (October) | Total (November) |
|---|---|---|---|---|---|---|
| Nagqu | 254,298 | 76,819 | 1,417,534 | 2,539,949 | 1,671,832 | 2,616,768 |
| Amdo | 115,999 | 44,039 | 183,397 | 357,656 | 299,396 | 401,696 |
| Sog | 55,206 | 12,857 | 156,549 | 273,331 | 211,755 | 286,188 |
| Biru | 48,248 | 13,979 | 127,454 | 234,880 | 175,702 | 248,859 |
| Baqen | 23,458 | 6036 | 87,296 | 151,993 | 110,755 | 158,029 |
| Baingoin | 10,171 | 3000 | 66,831 | 110,719 | 77,002 | 113,719 |
| Nyainrong | 12,024 | 2834 | 63,565 | 107,115 | 75,589 | 109,949 |
| Nyima | 36,396 | 13,299 | 22,053 | 47,607 | 58,449 | 60,906 |
| Lhari | 6175 | 2064 | 37,994 | 78,866 | 44,169 | 80,931 |
| Xainza | 7865 | 2364 | 23,002 | 41,691 | 30,867 | 44,054 |
| Shuanghu | 9812 | 3784 | 3609 | 6462 | 13,421 | 10,246 |
| Mean | 52,695.65 | 16,461.32 | 199,025.90 | 359,115.41 | 251,721.54 | 375,576.74 |
| Pair mean | 36,234.326 | | −160,089.52 | | −123,855.192 | |
| t | 2.345 | | −1.643 | | −1.498 | |
| df | 10 | | 10 | | 10 | |
| Sig. (2-tailed) | 0.041 | | 0.131 | | 0.165 | |

df = Degrees of freedom.

Table 3 reveals the number of tourist inflows and outflows and the total number of tourist flows in the Xigaze prefecture. The number of total tourist flow in the Xigaze prefecture indicates that the Samzhubze district has the highest tourist flow, whereas the Gamba district has the lowest number of tourist flows. In addition, the Samzhubze district acts as a central place connecting all the other districts in the Xigaze prefecture. Moreover, the tourist flows in the eastern and western parts are relatively the same. Through paired sample t-test, significant differences ($p = 0.017$) are found for tourist outflows in the Xigaze prefecture between October 2018 and November 2018. They are significant at 95% confidence. Regarding the total number of tourist flows of the Xigaze prefecture, the finding shows that the $p$ value is 0.062 and is significant at the 90% confidence interval. Specifically, tourist outflows and the total number of tourist flows of the Xigaze prefecture in November 2018 is greater than those in October 2018.

With the Tingri district as a center, tourist flows in the Xigaze prefecture are generally reflected by the primary flows among core nodes, the tertiary flows between the core and the minor nodes, and the normal moves among minor nodes. Thus, the Lhaze, Samzhubze, Bainang, and Tingri districts are the centers in western, central, eastern, and southern Xigaze, respectively. In summary, the tourist flow in the Xigaze prefecture is indicated by the structures of "balanced diffusion from the main center or diffusion from multiple centers." Although Lew and McKercher [4] indicated that certain tourists prefer time-efficient travel routes, the number of and the attractiveness of the attractions can effectively encourage tourists to extend their exploration in a certain region to achieve a balanced tourism development in different parts of the prefecture.

Table 4 indicates tourist inflows and outflows and the total number of tourist flows in the Lhasa prefecture. The Chengguan district in the Lhasa prefecture can be regarded as a distribution center for tourists, whereas the Nyemo district receives the least tourists among all areas in the Lhasa prefecture. Moreover, tourist flows are concentrated in the Chengguan and Doilungdeqen districts and are scattered in three different directions (i.e., north, west, and eastern parts). Damxung, Quxu, and Maizhokunggar districts represent the North, West, and East, respectively. A paired sample t-test shows that significant differences ($p = 0.094$) exist between October 2018 and November 2018 regarding tourist outflows of the Lhasa prefecture. They are significant at the 90% confidence interval. Specifically, the number of tourist outflows of the Lhasa prefecture in November 2018 is greater than that in October 2018.

**Table 3.** Tourist flow in Xigaze prefecture.

| Xigaze | Inflows (October) | Inflows (November) | Outflows (October) | Outflows (November) | Total (October) | Total (November) |
|---|---|---|---|---|---|---|
| Samzhubze | 1,448,378 | 820,002 | 1,041,201 | 1,587,730 | 2,489,579 | 2,407,731 |
| Gyangze | 155,394 | 112,849 | 390,762 | 602,498 | 546,156 | 715,348 |
| Yadong | 90,274 | 61,572 | 197,020 | 380,285 | 287,294 | 441,856 |
| Lhaze | 86,163 | 50,927 | 219,331 | 293,309 | 305,494 | 344,237 |
| Bainang | 41,454 | 30,057 | 247,135 | 312,618 | 288,589 | 342,675 |
| Ngamring | 52,027 | 27,039 | 165,047 | 289,523 | 217,074 | 316,562 |
| Sa'gya | 75,691 | 50,271 | 138,439 | 179,724 | 214,130 | 229,995 |
| Namling | 103,507 | 94,099 | 65,544 | 91,882 | 169,050 | 185,981 |
| Kangmar | 27,369 | 19,342 | 75,425 | 124,151 | 102,794 | 143,493 |
| Tingri | 102,664 | 55,219 | 43,121 | 67,053 | 145,785 | 122,272 |
| Xaitongmoin | 56,035 | 36,641 | 46,816 | 75,090 | 102,850 | 111,731 |
| Rinbung | 31,170 | 25,562 | 63,130 | 81,907 | 94,300 | 107,469 |
| Saga | 56,758 | 30,050 | 31,673 | 59,683 | 88,431 | 89,734 |
| Gyirong | 38,491 | 29,420 | 20,964 | 34,365 | 59,456 | 63,785 |
| Nyalam | 42,411 | 28,571 | 15,654 | 25,658 | 58,064 | 54,229 |
| Dinggye | 23,477 | 17,139 | 10,214 | 20,771 | 33,691 | 37,911 |
| Zhongba | 30,667 | 15,251 | 9223 | 21,574 | 39,890 | 36,825 |
| Gamba | 7819 | 9044 | 3646 | 6077 | 11,465 | 15,121 |
| Mean | 137,208.12 | 84,058.64 | 154,685.84 | 236,327.68 | 291,893.96 | 320,386.33 |
| Pair mean | 53,149.473 | | −81,641.843 | | −28,492.370 | |
| t | 1.564 | | −2.651 | | −2.002 | |
| df | 17 | | 17 | | 17 | |
| Sig. (2-tailed) | 0.136 | | 0.017 | | 0.062 | |

df = Degrees of freedom.

**Table 4.** Tourist flow in Lhasa prefecture.

| Lhasa | Inflows (October) | Inflows (November) | Outflows (October) | Outflows (November) | Total (October) | Total (November) |
|---|---|---|---|---|---|---|
| Chengguan | 6,597,353 | 3,557,333 | 3,557,333 | 5,466,361 | 10,154,686 | 8,743,031 |
| Doilungdeqen | 2,065,976 | 1,561,062 | 1,561,062 | 2,659,161 | 3,627,038 | 4,108,468 |
| Damxung | 236,443 | 373,228 | 373,228 | 592,174 | 609,671 | 720,842 |
| Quxu | 283,617 | 240,554 | 240,554 | 427,819 | 524,171 | 686,374 |
| Dagze | 294,275 | 98,970 | 98,970 | 189,459 | 393,245 | 474,621 |
| Maizhokunggar | 219,974 | 143,443 | 143,443 | 264,134 | 363,417 | 392,683 |
| Lhunzhub | 100,884 | 102,913 | 102,913 | 172,585 | 203,797 | 253,700 |
| Nyemo | 35,831 | 67,411 | 67,411 | 91,875 | 103,241 | 138,681 |
| Mean | 1,229,294.12 | 706,853.72 | 768,114.12 | 1,232,946.21 | 1,997,408.24 | 1,939,799.92 |
| Pair mean | 522,440.408 | | −464,832.089 | | 57,608.319 | |
| t | 1.286 | | −1.935 | | 0.288 | |
| df | 7 | | 7 | | 7 | |
| Sig. (2-tailed) | 0.239 | | 0.094 | | 0.782 | |

df = Degrees of freedom.

Significant differences are generally found on the tourist outflows of the Lhasa prefecture, whereas no significant difference is found on tourist inflows of the Lhasa prefecture and the total number of tourist flows. In summary, the tourist flow of the Lhasa prefecture is reflected in the concentrated center with scattering in different directions. In other words, scenic tourist spots (i.e., Potala Palace and Jokhang Temple) in the city center play dominant roles in influencing the overall tourist flow. The tourist flow pattern in the Lhasa prefecture is indicated by the structure of "diffusion from the main center." The central tourist flow is similar to the findings of Leung et al. [24], that tourist activities were within the center area of Beijing during the Olympics in Beijing in 2008. However, the tourist flow pattern in Tibet has further expanded in different directions.

Table 5 reveals the number of tourist inflows, outflows, and the total number of tourist flows in the Shannan prefecture. In general, the Nedong district is a central place connecting the western, southern, and eastern parts of the Nyingchi prefecture. This district also has the most tourist flows, whereas Comai has the lowest number of tourist flows. Furthermore, Nedong and its nearby districts receive more tourist flow, whereas the western, eastern, and southern parts receive less tourist flow. Paired sample t-test shows significant differences ($p = 0.020$) between tourist outflows and the total number of tourist flows ($p = 0.015$) of the Shannan prefecture from October to November 2018. They are significant at the 95% confidence interval. Specifically, tourist outflows and the total number of tourist flows in the Shannan prefecture in November 2018 is greater than that in October 2018.

**Table 5.** Tourist flow in Shannan prefecture.

| Shannan | Inflows (October) | Inflows (November) | Outflows (October) | Outflows (November) | Total (October) | Total (November) |
|---|---|---|---|---|---|---|
| Nedong | 587,336 | 315,646 | 516,000 | 866,457 | 1,103,335 | 1,182,103 |
| Gonggar | 151,083 | 149,303 | 360,413 | 586,290 | 511,496 | 735,593 |
| Chanang | 54,672 | 46,220 | 168,573 | 220,406 | 223,245 | 266,625 |
| Lhunze | 39,785 | 30,461 | 137,149 | 194,442 | 176,933 | 224,902 |
| Gyaca | 58,320 | 42,163 | 82,902 | 150,281 | 141,222 | 192,444 |
| Qusum | 25,255 | 15,318 | 115,866 | 204,711 | 141,121 | 220,029 |
| Nagarze | 22,909 | 20,361 | 111,297 | 147,569 | 134,206 | 167,930 |
| Sangri | 59,675 | 35,378 | 45,572 | 90,763 | 105,247 | 126,141 |
| Cona | 24,256 | 15,873 | 51,268 | 86,673 | 75,523 | 102,545 |
| Qonggyai | 14,712 | 8808 | 23,970 | 34,512 | 38,682 | 43,321 |
| Lhozhag | 17,452 | 10,457 | 3781 | 6193 | 21,233 | 16,650 |
| Comai | 14,366 | 12,853 | 3951 | 6433 | 18,317 | 19,286 |
| Mean | 89,151.55 | 58,570.06 | 135,061.78 | 216,227.44 | 224,213.32 | 274,797.50 |
| Pair mean | 30,581.486 | | −81,165.667 | | −50,584.181 | |
| t | 1.390 | | −2.716 | | −2.867 | |
| df | 11 | | 11 | | 11 | |
| Sig. (2-tailed) | 0.192 | | 0.020 | | 0.015 | |

df = Degrees of freedom.

In general, significant differences are found between tourist outflows and the total number of tourist flows. The tourist flow in the Shannan prefecture is generally indicated by the structures of "clustering from the main center" and "diffusion from a clustered circle." The identified tourist flow is considered a relatively balanced tourist flow, reflecting the primary flows between core nodes and the secondary flows scattered in different directions. Contemporary urban transportation models assume that the majority of people will take the most efficient route in a tourist destination if possible [4,30]. However, the findings of the present study reflect that a region may achieve such a balanced tourist flow by considering the convenience of transportation and the attractiveness of the attractions. Thus, transportation is considered an important factor affecting the spatial distribution of tourist flow [4,31].

Table 6 shows tourist inflows and outflows and the total number of tourist flows in the Nyingchi prefecture. The Nyingchi district in the Nyingchi prefecture is a central place that connects the areas in three different directions in the Nyingchi prefecture. The Nyingchi district has the highest tourist flow, whereas Nangxian has the lowest tourist flow. Furthermore, Nyingchi and its nearby districts receive more tourist flows than the districts that are remote to the Nyingchi district. In other words, districts that are far away from the Nyingchi district receive less tourist flows than nearby districts. The paired sample t-test shows significant differences ($p = 0.041$) in tourist outflows in the Nyingchi prefecture between October 2018 and November 2018. They are significant at the 90% confidence interval. Specifically, tourist outflows of the Nyingchi prefecture in November 2018 were more than those in October 2018.

**Table 6.** Tourist flow in Nyingchi prefecture.

| Nyingchi | Inflows (October) | Inflows (November) | Outflows (October) | Outflows (November) | Total (October) | Total (November) |
|---|---|---|---|---|---|---|
| Nyingchi | 1,379,217 | 709,339 | 768,614 | 1,243,659 | 2,147,831 | 1,952,998 |
| Mainling | 295,975 | 176,234 | 278,128 | 442,883 | 574,103 | 619,117 |
| Bome | 153,235 | 86,128 | 169,398 | 278,394 | 322,633 | 364,523 |
| Gongbo'gyamda | 171,079 | 72,965 | 123,301 | 220,892 | 294,380 | 293,857 |
| Zayu | 70,419 | 38,813 | 48,876 | 95,948 | 119,295 | 134,761 |
| Nangxian | 84,290 | 43,960 | 43,405 | 75,639 | 127,695 | 119,598 |
| Mean | 359,035.97 | 187,906.56 | 238,620.26 | 392,902.43 | 597,656.23 | 580,808.99 |
| Pair mean | 171,129.41 | | −154,282.16 | | 16,847.24 | |
| t | 1.700 | | −2.303 | | 0.459 | |
| df | 5 | | 5 | | 5 | |
| Sig. (2-tailed) | 0.150 | | 0.070 | | 0.665 | |

df = Degrees of freedom.

Tourist flow is generally concentrated in the Nyingchi district, and the flow is scattered in the western and eastern parts. Gongbo'gyamda and Nangxian districts represent the western direction, and the Zayu district indicates the eastern direction. In contrast to other prefectures, tourist flow in the Nyingchi prefecture is different in attracting more tourist flow in the western part than that in the eastern part. Similar to the tourist flow pattern identified in the Lhasa prefecture, the tourist flow pattern in the Nyingchi prefecture is also indicated by the structure of "diffusion from the main center" but with less tourist flow directions.

In conclusion, from a regional perspective, tourist flow in the western parts is lower than that in the eastern parts of Tibet. Ma and Wu found that the spatial structure of a destination is not in a state of equilibrium, and tourists tend to prefer the products in the eastern part of Xi'an, China [3]. The total number of tourist flows in different prefectures in November is generally more than that in October 2018. Jin et al. [10] stated that the tourist flow pattern is characterized by "diffusion from a single center," "clustering to a single center," and "balancing between multiple centers." Findings reveal that the tourist flow patterns of the Nagari, Lhasa, and Nyingchi prefectures mainly belong to the "(diverse) diffusion from the main center." The tourist flow pattern in Nagqu prefecture extends the identified tourist flow pattern by adding "clustering to the main center" to "diffusion from the main center." By contrast, Xigaze and Shannan prefectures reflect different tourist flow patterns despite what is identified by previous studies. The tourist flow pattern of the Xigaze prefecture is indicated by "a balanced diffusion from the main center or balancing between multiple centers," and that of the Shannan prefecture is reflected by "diffusion from a clustered circle." Furthermore, the tourist flow patterns in different prefectures in Tibet are characterized by primary, secondary, and tertiary flows [10].

## 5. Implications and Conclusions

The present study uses the SNA theory to visualize tourist flow and specifically examine tourist inflows, outflows, and the total number of tourist flow, thereby identifying tourist flow patterns in each of the different prefectures in Tibet. The findings show that the Lhasa prefecture has the most tourist flow among other prefectures in Tibet. Specifically, the Lhasa prefecture attracts the largest number of tourist flow, followed by Nagqu, Xigaze, Nyingchi, Shannan, and Ngari prefectures. Similar to the concept of distance decay [32,33], the findings of the present study reveal that distance also plays a vital role in determining the amount of tourist flow in Tibet. The overall structure of the tourist flow pattern is spreading from the center to outer parts, and the tourist flow in eastern parts is stronger than that in western parts. Zhong et al. detected regional disparity and found that China's eastern economic belt continues to have tourism-related benefits [8]. The present study extends SNA by integrating tourist flow into the movement patterns of tourists to identify tourist flow pattern. The findings of the present study not only provide an overall picture of the tourist flow in a certain region (i.e., Tibet) but also indicate the detailed tourist flow pattern in each of the prefectures in Tibet. Furthermore, they contribute to the literature by providing tourist flow pattern from a regional perspective and extending the identification of the structures of tourist flow pattern identified by previous studies.

The findings of the present study also provide valuable practical implications to tourism practitioners regarding the infrastructure construction of a certain region. Becken et al. [2] pointed out that the information about international visitor arrivals to New Zealand can provide sufficient information at a geographic level for infrastructure-related decision-making. The findings suggest that the Tibet tourism bureau must consider increasing the tourist flow in the western part to balance the development between eastern and western parts. The Ngari prefecture has the lowest tourist flow. Thus, tourist practitioners must come up with corresponding measures, such as infrastructure construction and transportation consideration, to attract more tourists. Among all different prefectures in Tibet, the Shannan and Xigaze prefectures reflect a relatively balanced tourist flow that helps promote healthy and sustainable tourism development. In other words, "balancing between multiple centers" can be considered to facilitate the balanced tourism and economic development of different areas in a region.

In conclusion, although previous studies have identified either movement patterns or itineraries of tourists [4,5], considerations of tourist inflows and outflows are lacking. Hence, the present study identifies the inflows and outflows of tourists in Tibet based on the SNA to provide implications to balance its regional economic development and promote its sustainable tourism development. Tourist flows in different prefectures in Tibet are identified and analyzed by retrieving data generated by the mobile phone signal of China Unicom. The findings show that the Lhasa prefecture is the transportation hub of Tibet. Tourist flow in the eastern part is generally stronger than that of the western part in Tibet. The tourist flow pattern identified for different prefectures in Tibet mainly includes "(diverse or balanced) diffusion from the main center," "clustering to the main center," and "diffusion from a clustered circle." In addition, future studies can be extended to other countries and regions to investigate tourist flow patterns to promote sustainable development by balancing regional economic development. The present study has three limitations. First, positioning-related errors may exist through tracking tourist flow by a mobile signal. In addition, the present study only tracked flow patterns of tourists who used the China Unicom telecommunication service, but those tourists who used other mobile telecommunication companies were not tracked. Moreover, the present study only investigated tourist flow patterns in each of the prefectures in Tibet, and tourist flow patterns that cross different prefectures were not considered. Hence, future research can track the flow patterns of tourists who use different mobile telecommunication companies and compare the differences in tourist flow patterns who use different mobile telecommunication companies. Future studies can further explore the different preferences of tourists from different countries or origins and examine tourist flow patterns that cross regions to provide accurate implications for tourism practitioners regarding regional tourism development.

**Author Contributions:** Conceptualization, L.Z.; methodology, L.Z.; software, R.L.; formal analysis, R.L.; writing—original draft preparation, L.Z.; writing—review and editing, S.S. and L.Y.; supervision, S.S.; funding acquisition, L.Z. All authors have read and agreed to the published version of the manuscript.

**Funding:** This research was funded by National Natural Science Foundation of China (Grant No. 71673015); Beijing Social Science Fund (No. 19 JDXCA0059), Ethnic research project of the National Committee of the People's Republic of China (2020-GMD-089); and the Conference Funding Subsidy of Ritsumeikan Asia Pacific University.

**Conflicts of Interest:** The authors declare that they have no conflict of interest.

## References

1. Degen, W. The Influence of Beijing-Shanghai High-speed Railway on Tourist Flow and Time-Space Distribution. *Tour. Trib. Lvyou Xuekan* **2014**, *29*, 75–82.
2. Becken, S.; Vuletich, S.; Campbell, S. Developing a GIS-supported tourist flow model for New Zealand. In *Developments in Tourism Research*; Routledge: New York, NY, USA, 2007; pp. 107–121.

3. Ma, X.; Wu, B. Spatial Structure of Tourist Flow in Xi'an Tourism Region. *Geogr. Geo-Inf. Sci.* **2004**, *5*, 95–97.
4. Lew, A.; McKercher, B. Modeling Tourist Movements. *Ann. Tour. Res.* **2006**, *33*, 403–423. [CrossRef]
5. Zhong, L.; Sun, S.; Law, R. Movement patterns of tourists. *Tour. Manag.* **2019**, *75*, 318–322. [CrossRef]
6. Cox, K.R. *Man, Location, and Behavior: An Introduction to Human Geography*; John Wiley & Sons: New York, NY, USA, 1972.
7. Oppermann, M. A model of travel itineraries. *J. Travel Res.* **1995**, *33*, 57–61. [CrossRef]
8. Zhong, S.; Zhang, J.; Li, X. A Reformulated Directional Bias of Tourist Flow. *Tour. Geogr.* **2011**, *13*, 129–147. [CrossRef]
9. Liu, F.J.; Zhang, J.; Zhang, J.H.; Chen, D.D. Roles and functions of provincial destinations in Chinese inbound tourist flow network. *Geogr. Res.* **2010**, *6*, 1141–1152.
10. Jin, C.; Xu, J.; Huang, Z.; Cao, F. Analyzing the characteristics of tourist flows between the scenic spots in inner city based on tourism strategies: A case study in Nanjing. *J. Geogr. Sci.* **2014**, *69*, 1858–1870. [CrossRef]
11. Jansen-Verbeke, M.; Spee, R. A regional analysis of tourist flows within Europe. *Tour. Manag.* **1995**, *16*, 73–80. [CrossRef]
12. Zhang, J.; Zhang, Y. Trade-offs between sustainable tourism development goals: An analysis of Tibet (China). *Sustain. Dev.* **2019**, *27*, 109–117. [CrossRef]
13. Xinhua News. "Tibet Winter Travel"—The Number of Tourists to Tibet Exceeded 30 Million the First Time. Available online: http://www.xinhuanet.com/local/2019-01/10/c_1123972865.htm (accessed on 15 May 2020).
14. Ahas, R.; Aasa, A.; Mark, Ü.; Pae, T.; Kull, A. Seasonal tourism spaces in Estonia: Case study with mobile positioning data. *Tour. Manag.* **2008**, *28*, 898–910. [CrossRef]
15. Buckley, P.J.; Mirza, H.; Witt, S.F. Japan's international tourism in the context of Its international economic relations. *Serv. Ind. J.* **1989**, *9*, 357–383. [CrossRef]
16. Kulendran, N. Modelling quarterly tourist flows to Australia using cointegration analysis. *Tour. Econ.* **1996**, *2*, 203–222. [CrossRef]
17. Coshall, J. Spectral analysis of international tourism flows. *Ann. Tour. Res.* **2000**, *27*, 577–589. [CrossRef]
18. Dimoska, T.; Petrevska, B. *Tourism and Economic Development in Macedonia, Conference Proceedings Tourism & Hospitality Industry 2012*; Univerity of Rijeka, Faculty of Tourism and Hospitality Management: Opatija, Croatia, 2012; pp. 12–20.
19. Patuelli, R.; Mussoni, M.; Candela, G. The effects of World Heritage Sites on domestic tourism: A spatial interaction model for Italy. *J. Geogr. Syst.* **2013**, *15*, 369–402. [CrossRef]
20. Wasserman, S.; Faust, K. *Social Network Analysis: Methods and Applications*; Cambridge University Press: Cambridge, UK, 1994.
21. Scott, J.; Tallia, A.; Crosson, J.C.; Orzano, A.J.; Stroebel, C.; DiCicco-Bloom, B.; O'Malley, D.; Shaw, E.; Crabtree, B. Social network analysis as an analytic tool for interaction patterns in primary care practices. *Ann. Fam. Med.* **2005**, *3*, 443–448. [CrossRef]
22. Otte, E.; Rousseau, R. Social network analysis: A powerful strategy, also for the information sciences. *J. Inf. Sci.* **2002**, *28*, 441–453. [CrossRef]
23. Shih, H.-Y. Network characteristics of drive tourism destinations: An application of network analysis in tourism. *Tour. Manag.* **2006**, *27*, 1029–1039. [CrossRef]
24. Leung, X.Y.; Wang, F.; Wu, B.; Bai, B.; Stahura, K.A.; Xie, Z. A social network analysis of overseas tourist movement patterns in Beijing: The impact of the Olympic Games. *Int. J. Tour. Res.* **2012**, *14*, 469–484. [CrossRef]
25. Raisi, H.; Baggio, R.; Barratt-Pugh, L.; Willson, G. Hyperlink network analysis of a tourism destination. *J. Travel Res.* **2018**, *57*, 671–686. [CrossRef]
26. David-Negre, T.; Hernández, J.M.; Moreno-Gil, S. Understanding tourists' leisure expenditure at the destination: A social network analysis. *J. Travel Tour. Mark.* **2018**, *35*, 922–937. [CrossRef]
27. Page, S. *Transport and Tourism*, 2nd ed.; Prentice Hall: Harlow, UK, 1999.
28. Zeng, B. Pattern of Chinese tourist flows in Japan: A Social Network Analysis perspective. *Tour. Geogr.* **2018**, *20*, 810–832. [CrossRef]
29. Toha, M.A.M.; Ismail, H.N. A heritage tourism and tourist flow pattern: A perspective on traditional versus modern technologies in tracking the tourists. *Int. J. Built Environ. Sustain.* **2015**, *2*. [CrossRef]
30. Meyer, M.D.; Miller, E.J. *Urban Transportation Planning: A Decision-Oriented Approach*; McGraw-Hill: New York, NY, USA, 1984.

31. Wang, D.; Chen, T.; Lu, L.; Wang, L.; Alan, A.L. Mechanism and HSR effect of spatial structure of regional tourist flow: Case study of Beijing-Shanghai HSR in China. *Dili Xuebao/Acta Geogr. Sin.* **2015**, *70*, 214–233.
32. Lee, H.A.; Guillet, B.D.; Law, R.; Leung, R. Robustness of distance decay for international pleasure travelers: A longitudinal approach. *Int. J. Tour. Res.* **2012**, *14*, 409–420. [CrossRef]
33. Mckercher, B.; Lew, A.A. Distance decay and the impact of effective tourism exclusion zones on international travel flows. *J. Travel Res.* **2003**, *42*, 159–165. [CrossRef]

MDPI
St. Alban-Anlage 66
4052 Basel
Switzerland
Tel. +41 61 683 77 34
Fax +41 61 302 89 18
www.mdpi.com

*Sustainability* Editorial Office
E-mail: sustainability@mdpi.com
www.mdpi.com/journal/sustainability

www.ingramcontent.com/pod-product-compliance
Lightning Source LLC
LaVergne TN
LVHW070623170726
843515LV00004B/161

* 9 7 8 3 0 3 6 5 0 4 6 0 5 *